Engineering Aspects of Howrah Bridge at Kolkata (1943)

Engineering Aspects of Howrah Bridge at Kolkata (1943)

Amitabha Ghoshal

CRC Press
Taylor & Francis Group
Boca Raton London New York

CRC Press is an imprint of the
Taylor & Francis Group, an **informa** business

First edition published 2021
by CRC Press
6000 Broken Sound Parkway NW, Suite 300, Boca Raton, FL 33487-2742

and by CRC Press
2 Park Square, Milton Park, Abingdon, Oxon, OX14 4RN

© 2021 Amitabha Ghoshal
CRC Press is an imprint of Taylor & Francis Group, LLC

Library of Congress Cataloging-in-Publication Data
Names: Ghoshal, Amitabha, author.
Title: Engineering aspects of Howrah Bridge at Kolkata (1943) / Amitabha Ghoshal.
Description: First edition. | Boca Raton, FL : CRC Press, 2021. | Includes bibliographical references and index.
Identifiers: LCCN 2021007642 (print) | LCCN 2021007643 (ebook) | ISBN 9780367544744 (hardback) | ISBN 9780367544799 (paperback) | ISBN 9781003089438 (ebook)
Subjects: LCSH: Howrah Bridge (Howrah and Kolkata, India)--History.
Classification: LCC TG104.K65 G46 2021 (print) | LCC TG104.K65 (ebook) | DDC 624.2/19095414--dc23
LC record available at https://lccn.loc.gov/2021007642
LC ebook record available at https://lccn.loc.gov/2021007643

ISBN: 978-0-367-54474-4 (hbk)
ISBN: 978-0-367-54479-9 (pbk)
ISBN: 978-1-003-08943-8 (ebk)

Typeset in Times
by SPi Global, India

This book is dedicated to the memory of the Indian workmen who worked through rain and sun, at dizzy heights, to improve the daily lives of city dwellers; to men like Omeruddin Khan and Harbans Singh, who mastered the skill and contributed to successful completion of this and many more large bridges in India.

Contents

Preface

Howrah Bridge had been a part of my weekend commute from Bengal Engineering College (BE College), via Howrah to my residence in Calcutta for the four years that I spent in the college hostel. Every trip across the bridge inspired me to dream of building bridges like it!

It was not just an inspiration but also my instructor! Later, when I received an opportunity to be a part of the successful tender design process for the next bridge about a kilometer downstream (Vidyasagar Setu), I studied and critically scrutinized all the published literature about how the old bridge was designed and built. I was fortunate to have been apprenticed, and later employed, by the BBJ Construction Co. (BBJ), the Calcutta-based company that was involved in construction of Howrah Bridge. BBJ subsequently went on to build almost all the major bridges using steelwork in the country. They were also the EPC contractor for superstructure of the Vidyasagar Setu, the first major cable-stayed bridge in the country, at that time planned as the longest cable-stayed span in the world.

The narratives of Howrah Bridge, as recorded by the engineers responsible for the bridge, fascinated me, and when I was requested by the publishers to write an engineering book my immediate first choice was the story of Howrah Bridge. The task proved more challenging than I had thought, but my publishers were kind and patient in giving me sufficient time to complete the book.

The bridge did more than allowing people of the twin city to cross this mighty river, a branch of Holy Ganges, revered and loved by the Indians as a part of their history and mythology alike. It made the engineering community across the world take note of the innovations that went into the bridge, and the Indians proud that the bridge could be built by Indian workmen and engineers, with most materials sourced from within the country, supervised by a small group of bridge engineers from Britain who were highly ranked in their respective fields of specialization. Constructed in a very short period, much of which was during the Second World War when logistics everywhere were severely strained, the bridge stands as a tribute to the meticulous planning and foresight of the experts in charge. Built across a busy river and in a congested area of the city, construction work did not have any unplanned delay nor any accident worth mentioning. It is a story I found worth recording!

The bridge made a lasting impact on the construction industry in India. Built only three years before India earned independence following two hundred years of colonial rule, the experience gained in this bridge by Indian engineers and workmen alike stood the country in good stead, when the country had to depend entirely on its own for the development programs. The experience gained in the drawing offices of fabricators on the detailing of such major steelwork, and the fabrication skills attained in the shops, helped the country to design and erect many major steel bridges in the country immediately afterwards. In fact, Indians never required foreign help for any major bridgework afterwards. The confidence created amongst workmen in cantilever construction has been sustained and most bridges in India still use this technique for bridges across major rivers and valleys.

It is not surprising that this bridge has captured the imagination of generations of Indian engineers. For the residents of Calcutta, the bridge is a defining landmark of their beloved city and their identity!

It is my fervent hope that you dear readers, like me, will find the story of this bridge inspirational.

Acknowledgement

Writing a first book, as I discovered, is a strenuous task, and I sought help from my close circle without any hesitation. It is only fair that I record my grateful thanks in text!

The organisations that nurtured my professional development through the years, The Bengal Engineering College (now IIEST, a centre of excellence for technical education) tops the list, followed by The Braithwaite Burn and Jessop Construction Co Ltd (BBJ), employer for first 21 years of my career, and intimately connected with the construction of Howrah Bridge, and the STUP Consultants Pvt Ltd, with whom I was associated for 40 years in different roles, and from which I spread my professional footprint to more than 20 countries, are acknowledged with reverence.

Institution of Civil Engineers, U.K., (ICE) are the greatest source of heritage civil engineering structures in British India, and their publications form the backbone of this book. Permission to source from their publications has been generously granted by officials of ICE, an organisation whose Fellowship I have treasured for more than 50 years.

My friend and professional associate for 65 years, K.V. Narasimhan, who helped me by reading and editing every word written for the book, has been a great support, a nonagenarian, mentally fit as a fiddle.

My professional colleagues, Anirban Sengupta, Sumantra Sengupta, Achyut Ghosh, Chandan Ganguly and Sundar Bannerji, who joined me in celebrating 75 years of this unique structure by holding a day long Seminar in 2018, have been source of many information bites.

Colleagues Soumitra Ghosh and Aniruddha Basu Majumder have helped untiringly in giving shape to the final product.

My intimate family members - my sons Subhasis and Soumya, their wives Padma and Anjana and grandchildren Ila and Uma kept on encouraging me whenever the enthusiasm to complete the task flagged.

My wife Sila has been generously allowing me to ignore family responsibilities and devote myself to the professional pastimes that I kept on accruing as opportunities came, and without this freedom it would not have been possible to do whatever I could for the profession that sustained me!

The publication team of CRC Press have been sympathetically indulgent during the difficult days of COVID-19 pandemic and my grateful acknowledgement to them.

Introduction

1

THIS BOOK

This book is all about the New Howrah Bridge, one of the great bridges of the world and the icon of Calcutta, one of the great cities of the world. The book describes the fascinating story of the evolution of the New Howrah Bridge, how it was conceived to cater to the needs of the people, the condition of the city and the construction challenges (Figure 1.1).

The most important thing about major bridges is perhaps not their size, the innovations, the steel and concrete involved, or the engineering skill that has gone into them but that they connect people and create bonds between them, and because of this vital function the bridge in the course of time becomes identified with the very people it has connected. The bridges become the icons of the cities of their location. Thus cities/people and the bridges become inextricably linked. San Francisco and the Golden Gate, London and the Tower Bridge, Calcutta and the Howrah Bridge are the cases in the point.

The port city of Calcutta was built on the eastern bank of the mighty river away from the mainland of India, perhaps by design to avoid marauding bands roaming around during the decline of the Mughal empire. The river provided a natural protection, and that advantage was found to be more important than the health challenges posed by the poorly drained, marshy, unhealthy areas that were the main feature of the terrain on the eastern bank.

THE CRYING NEED FOR A BRIDGE

It is 1850 AD. Calcutta had already been the capital for East India Company for 78 years, starting in 1772 when Warren Hastings moved in from Murshidabad. The "Sepoy Mutiny" was seven years into the future and the Crown would be

FIGURE 1.1   Shining in day time

taking over the possessions and assets of the East India Company one year later, in 1858.

On the eastern bank of the mighty river, the three villages, which Job Charnok was presumed to have bought from local landlords, had already grown into a big city with a burgeoning population of about 230,000.

Calcutta was already a bustling trade center due to the rapid development of British private shipping. The hinterland for the port was very large – eastern areas of what is now Uttar Pradesh, south-western, southern and all eastern areas of India. The import–export trade with Britain was expanding very quickly.

The flourishing textile industry and a booming economy drew investments in railways from private parties. Howrah station was opened, and the first train left in 1854. (British-owned private railways had started operating in India and the trains were running from 1853.) The terminals of the Eastern Bengal Railway, serving the eastern part of northern India, and Bengal Nagpur Railway, serving the south and eastern part of the country, were both based in Howrah, on the western bank of the river, directly across Calcutta.

The railway terminal at Howrah brought in the goods for the fast-growing port and cleared the goods imported through Calcutta Port. There was also very heavy passenger movement between Howrah and Calcutta. But the mighty river was in between, the absence of a bridge being a great barrier preventing easy and smooth flow of goods and men.

The need for a land corridor grew and demands for a bridge across the river became strident. A committee was formed in 1855–1856 to review alternatives for constructing a bridge across the river. Matters languished and, in 1862, the government asked George Turnbull, Chief Engineer of Eastern Bengal Railway, to study the feasibility of a bridge across the Hooghly River. He submitted a report in March recommending a bridge 12 miles north of Calcutta rather than at the required location because of technical and cost considerations.

THE PONTOON PREDECESSOR

The Calcutta Port Trust was formed in 1870 and the Government of Bengal passed the Howrah Bridge Act in 1871 empowering the Lieutenant Governor to have the bridge constructed with government funds under the aegis of the port commissioners (Ward & Bateson, 1947).

Finally, the responsibility fell on the fabled engineer Bradford Leslie (later Sir Bradford KCIE), Chief Engineer to the Eastern Bengal Railway, for the construction of a pontoon bridge across the Hooghly between Howrah on the west bank and Calcutta on the east bank. This floating bridge was the (first) Howrah Bridge.

Sir Leslie was a pupil of the legendary Isambard Kingdom Brunel and was a famous bridge engineer in his own right. He went on to design the double track railway bridge, again across Hooghly, 39 km (24 miles) upstream of Howrah Bridge. That artistically shaped balanced cantilever bridge, called Jubilee Bridge, was opened in 1887, and served with speed restrictions until 2016, when a new bridge was commissioned.

The floating Howrah Bridge was of a total length of 1528 ft (466 m), between abutments, and had a road carriageway of 48 ft (14.6 m) and two 7 ft (2.1 m) footpaths. It was fabricated in England, shipped to Calcutta in parts and erected at the site. During erection, the bridge was damaged due to the great cyclone of 20 March 1874, when a steamer, *Egaria*, broke from its moorings and sank three pontoons of the bridge, damaging 200 ft of the bridge.

In spite of this, the pontoon bridge was opened to traffic in October 1874. It was initially designed for a life of 25 years, but it had to, and did, function for 69 years until the New Howrah Bridge could be opened to traffic in 1943.

The design of this floating bridge had to cater to the intrinsic challenges provided by the Hooghly River, which were having a flat terrain on both banks and tidal variations of up to 20 ft (6.1 m). The traffic across the bridge consisted predominantly of handcarts carrying goods and bullock-driven carts ferrying cargo, apart from a very large number of pedestrians disgorged by incoming trains and wending their way to reach the business center of Calcutta, a kilometer downstream on the Calcutta bank. Provision had to be made to cater for the substantial river traffic that carried export cargo to the port, located a few kilometers downstream, and carried back finished goods imported though the port.

The design of the pontoon provided for one central removable segment of 200 ft (61 m), an innovative arrangement at the time, for permitting large ships to move through to the repair facilities and graving docks located upstream on the banks. Two shore spans, one at each end, were provided with a headroom of 22 ft (6.7 m) for unhindered movement of smaller river crafts, by installing 160 ft (48.8 m) long bowstring girders, which were supported on pontoons at one end and the abutment on the other.

For tackling the tidal variation of more than 20 ft (6.1 m) in the river, the maximum gradients on the shore spans were kept at 1 in 16 up or down, during the extreme levels of tide in the river. The deck level on the central portion of the floating bridge was kept level, with a 1 in 40 slope at either end toward the shore spans (Figure 1.2).

With traffic on the bridge rapidly increasing with time, users faced a lot of problems. At times of extreme tides, the steep approach slopes were difficult to negotiate for animal-driven carts, causing traffic jams of long rows of vehicles at the approaches. Dense traffic-induced variations in level of up to 2 ft (0.6 m), between adjacent pontoons, made vehicle movement jerky. The removal of the central section for the passage of large ships, though normally once a day, depending on tide timings, created long rows of traffic at either approach where space was limited (Figure 1.3). The port commissioners, who were the

FIGURE 1.2   Howrah Bridge: supported on pontoons (Ward & Bateson 1947)

FIGURE 1.3   The traffic impasse

managing authorities for the river as well as the bridge, made improvements to the approaches and the bridge itself at regular intervals.

Let us hear from a man of the time, Mr. Bagley MICE, writing about the pontoon bridge, in the first issue of The Institution of Engineers' magazine in September 1921:

> The existing pontoon bridge was designed and built in 1874 by Mr. (now Sir) Bradford Leslie, then Chief Engineer of the East Indian Railway. It was sanctioned by the then Viceroy of India, Lord Mayo over the heads of all the experts who with scarcely an exception prophesied disaster from cyclones, and bores (tidal waves). It has fully justified the sagacity of the great proconsul and the genius of the designer.
>
> For 46 years it has been in continuous use, without accident or hitch of any kind and an inestimable boon to the public. **But its day is done.**

ORIGINS OF THE NEW HOWRAH BRIDGE

With the growth of the Calcutta Port and city and the rapid increase in rail goods and passenger traffic at Howrah, the pontoon bridge was rapidly becoming inadequate. All the stakeholders felt the pressure and arrangements for a long-term solution had to be set in motion (Ward & Bateson, 1947)

A 1906 committee, headed by R.S. Highet, Chief Engineer of the Eastern Bengal Railway, and W.B. McCabe, Chief Engineer of Calcutta Corporation, made a detailed study of the traffic and all relevant requirements that the new crossing had to meet. The obligatory conditions for the design included for:

> ➤ Unhindered/maximal movement of water crafts along the river
> ➤ Unhindered/maximal traffic movement along the carriageway of the bridge
> ➤ Carriageway suitable for six lanes of traffic and pedestrian path on both sides
> ➤ Minimum obstruction to hydraulic flow in the river
> ➤ Adequate clearance below the bridge structure for the river traffic to move during extreme tide conditions
> ➤ Maximum slope of 1 in 35 for the approaches, to allow animal-drawn vehicles to climb.

The committee considered the following six options:

1. Large ferry steamers capable of carrying vehicular loads
2. A transporter bridge
3. A tunnel
4. A bridge on piers
5. A floating bridge
6. An arch bridge.

The committee decided on a floating bridge. Tenders were floated and a prize money was declared for the firm whose design would be accepted.

Eighteen different designs were received from nine bidders in 1912, and the design submitted by the German firm MAN (Maschinenfabik Augsburg Nurnberg), which incorporated a Bascule-type girder for movement of ships, was selected and received the prize money. The construction of the bridge, however, could not commence due to the outbreak of the First World War (1914–1919). The old pontoon bridge had to be partially repaired and refurbished in 1916 and 1925 to be of service and to cope with the increasing traffic.

To quote Mr. Bagley again:

It is unnecessary at a meeting at Calcutta to dwell at any length upon the urgency of the case [the need for a new bridge – author] because it is evident to every traveler who arrives in Howrah or departs from it and falls victim to intolerable congestion on the existing bridge with its attendant delays and risks. At Howrah we have the terminus of the Railways bringing into the Metropolis the enormous traffic in passenger and goods from all the North, West and South of to a bottleneck forty-two feet wide and a quarter of mile in length sometimes with gradients which make wheeled traffic impossible. This state of things would be incredible if it were not demonstrated to thousands daily.

After the war, with the realization that a new bridge was imminently required, a new committee was appointed in 1921, chaired by an eminent engineer industrialist, Sir Rajendra Nath Mookerjee, with members nominated from the Port Commissioners and many internationally well-known engineers of that time. The Government of Bengal decided to refer the whole issue to the expert committee that came to be known as Mookerjee Committee. With this appointment, the work on the formulation of the new bridge started to take shape.

The committee commenced its work and collected data on navigational aspects of the river, soil condition at the site, the hydraulics of the river and other relevant matters as well as on local engineering capabilities.

Feasibility of the following types of bridges was examined:

➢ Single span arch
➢ Cable suspension
➢ Multiple pier and girder
➢ Floating
➢ Cantilever and balanced cantilever.

The committee gave its findings on the different bridge types in its report.

An arch bridge was not considered suitable as even at a good depth the clay and silt soil in the site would not be able to resist the horizontal thrust to which the foundations would be subjected.

A suspension bridge for a long span such as the present one needed anchors that had to resist large pulling forces, and the soil at the site would not be able to provide the required resistance.

A pier and girder bridge with smaller spans was the most economical of the options and easy to build, but the bridge would affect the scouring and silting pattern of the river and influence the navigability of the river, and this was not acceptable. The port authorities were averse to taking any risk on this account as the functioning of the port was of supreme importance.

A floating bridge offered an economical and fast solution for bridging, but it was considered a short-term solution even with removable sections provided. It was also considered vulnerable to damage by ships losing rudder control and a risk for both road and river traffic.

The committee's recommendation was for a cantilever bridge with suspended girders as the most desirable solution and the bridge configuration finally adopted in 1935 was close to the one recommended by the committee.

New Howrah Bridge Commissioners to the Government of Bengal had been set up in 1922 and the Mookerjee Committee submitted its report to the bridge commissioners.

In 1926, a legislation titled The New Howrah Bridge Act was passed and the commissioners for the port of Calcutta were made the commissioners for the new bridge.

In 1929, M/s. Rendel, Palmer and Tritton of London, the consultants retained by the port authorities, submitted their report and alternative estimates for two types of design, cantilever and floating, for the consideration of the authorities.

In 1930, the port commissioners recommended that a cantilever suspension type of bridge be considered, and that Messrs. Rendel, Palmer and Tritton (RPT) be entrusted with the design responsibilities.

After many false starts, the process of implementation of the bridge project did finally start with these actions.

REFERENCE

Ward, Arthur M. & Bateson, Ernest, *Paper No. 5601, Institution of Civil Engineers – The New Howrah Bridge, Calcutta: Design of the Structure, Foundations and Approaches,* 1947.

Planning of the Final Option

2

The terminal points for a major bridge in an urban setting have to be carefully selected. In this case there was the question of approaches sufficient for the specified grade; proper connectivity to obligatory points – Howrah Station/the then-business district/the port – being the more important ones; dispersal of traffic; likely need for demolition and/or relocation of existing structures on either side; extent of land acquisition if any and the difficulties/restrictions that it might pose. Harrison Road (now Mahatma Gandhi Road) was considered to be an important traffic dispersal corridor and ready connectivity with it was a requirement.

The pontoon bridge, being the lifeline between Howrah and Calcutta, had to be kept in proper shape and service until the new bridge was commissioned and opened to traffic.

The selected site was found to be satisfactory in respect of all these requirements. The alignment finalized was at right angles to both the banks, and was almost parallel to that of the pontoon bridge, which was downstream at points 630 ft (192 m) and 580 ft (177 m), respectively (Figure 2.1) (Ward & Bateson, 1947).

It was fortuitous that the land covering the approaches on both banks belonged to authorities and so there were no land acquisition issues. More importantly, there was enough land available for the specified gradient, which would provide welcome relief for the animal-driven carts.

CLEARANCES FOR RIVER TRAFFIC

Adequate clearance for ships plying underneath, and that for vessels moving upstream of the bridge to go to available repair facilities, was of utmost importance to the shipping port. A study indicated the requirement of this clearance

9

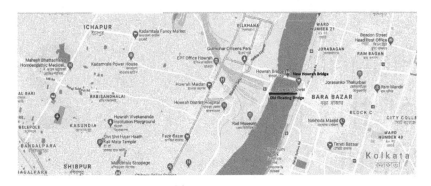

FIGURE 2.1   Alignment of new bridge, with respect to old floating bridge

to be about 29 ft (8.84 m) above the highest high flood level of the river at the bridge location.

The vertical clearance required above the deck level, for unencumbered movement of traffic on the bridge (the bridge was planned to carry tram cars also), was 19 ft (5.8 m).

To cater to the future growth of vehicular traffic (motorized trucks were already getting common), it was decided that the deck should have four lanes for vehicular traffic, and robust walkways, one on each side of the carriageway, to provide for the large number of commuters to the city.

Electric-driven tram cars had been plying from 1880 in the city and provision was made for one line of tram track each way, a very wise, environmentally significant decision!

The initial design had carriageways of 38 ft (11.6 m) – one each for up and down traffic – 29 ft (8.9 m) for vehicles and 9 ft (2.7 m) for a tram track – and a pedestrian path of 12 ft (3.66 m) on either side, a total of 100 ft (30.48 m).

The finally adopted design had a total carriageway width of 71 ft (21.6 m) and a pedestrian path of 15 ft (4.6 m) each on either side – closely similar to the original recommendation – but in between many alternative concepts had to be examined, to restrict the cost to be within the means of the government.

DEVELOPMENT OF FINAL DESIGN OF BRIDGE

The main span was agreed as 1500 ft (457 m) between the centers of towers, which was accepted by the port authorities as the safe hydraulic clearance and allowed for construction of the main tower foundations to be handled from the banks, a welcome situation for safe construction of the giant foundations.

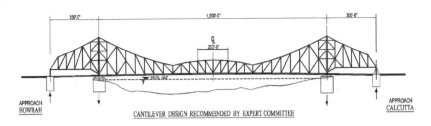

FIGURE 2.2 The profile of the bridge recommended by the Mookerjee Committee (Ward & Bateson, 1947)

The recommendation by the Mookerjee Committee was a 1500 ft (457 m) main span, and a balanced cantilever arrangement with suspended span at the center. The suspended span had a hog-backed profile, with a larger depth at center, to counter the maximum moment at that location. This, however, affected the aesthetics of the elevation (Figure 2.2) (Ward & Bateson, 1947).

The suggested span arrangement was for the deck to be at the bottom chord level, on a grillage of cross girders and stringers, supported on the bottom chord. The traffic had to take a turn for the exit in the anchor span stretch, as the area beyond the end anchor on Calcutta side was heavily built up. This was to be achieved by sloping the bottom chord up in the anchor span stretch and taking the deck out of the truss by modifying the bottom lateral bracings, before taking a right-angle turn to get on to the approach viaduct. A similar arrangement was envisaged for the Howrah end exit as well. It was later felt that this arrangement would become clumsy and need careful detailing.

Alternatives Explored

Balanced Cantilever Option

In 1922, the authorities decided to commission Sir Frederick Palmer, past president, Institution of Civil Engineers, UK, and a renowned consultant, to prepare a design of the bridge along with a cost estimate, for budgetary purposes, based on the balanced cantilever arrangement proposed by the Mookerjee Committee. The developed design of Sir Frederick, as shown in Figure 2.3, had some improved features.

The top chord of the truss was made a continuous concave curve, thereby improving the look of the elevation. The main span was retained at 1500 ft (457 m), but the anchor spans were increased to 446 ft (136 m), and the towers were made taller.

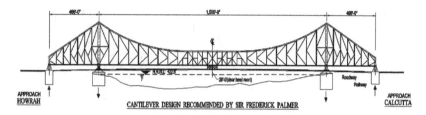

FIGURE 2.3   General view of the bridge conceived by Sir Frederick (1922) (Ward & Bateson, 1947)

The deck was proposed to be hung from the main truss, through hangers, from the bottom chord of the trussed girder, and the floor system thus remained independent of the main structural system, allowing seamless transition to the approach roads at either end. This arrangement gave the additional advantage of providing the main structure a much larger clearance of 64 ft (19.5 m) above the highest flood level-obviating any chance of damage to the bridge due to an accidental hit by an ocean-going vessel straying towards the bridge due to loss of its rudder control.

The estimated cost, worked out at £2.7 million, was way beyond the budget of the authorities, and it was decided to defer the project. The economy, after the First World War, was not in good shape, making such large capital investment challenging.

Floating Permanent Bridge

Due to the limitation of funds, in 1939 it was decided to re-examine the permanent floating bridge option though earlier this option had been rejected because of the vulnerability of such a bridge to accidental hits by vessels and the problems posed by the continuous changes in deck level with the tidal variation four times a day.

A floating bridge with five spans, supported on pontoons, was conceived as a non-opening bridge, as shown in Figure 2.4. Clearance of 37 ft (11.28 m) was provided in one span, at the center, for unrestricted transit of large vessels. Provision was made for a gradient of 1 in 40 at the approaches. The extreme gradients would change from 1 in 25 up during highest tide to 1 in 39 down during the lowest level of water.

The cost of this scheme, with carriageway and pedestrian way widths identical to Sir Frederick's design, came to £1.5 million, an amount that was attractive to the authorities.

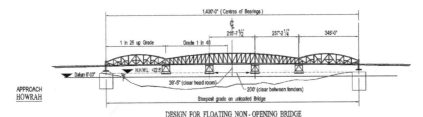

DESIGN FOR FLOATING NON-OPENING BRIDGE

FIGURE 2.4 Floating bridge arrangement with non-opening span (1929) (Ward & Bateson, 1947)

However, there was a lot of criticism of this proposal from many quarters, including the chambers of commerce, whose voice had gained in strength by that time.

This scheme had to be put on hold for the time being.

Modified Balance Cantilever Option

This was once again a review of the 1922 design, with some changes including the omission of tram lines, and reduced width of the footpaths. The cost reduction achieved with such cosmetic changes was marginal and not substantial enough.

Self-Anchored Suspension Bridge

By 1933 new exigencies arose with extensive deterioration of the pontoon bridge and concerns about a sudden collapse. A renewed search was started for a less costly alternatives. One solution examined was a self-anchored suspension bridge, where the problem of horizontal pull-on anchors were avoided by making the stiffening truss and deck girders carry the horizontal component of the tension in the cables supporting the long span. The carriageway width of the bridge was kept at 60 ft (18.2 m) and the span unchanged at 1500 ft (457 m).

The deck was hung from the stiffening girders to allow traffic to negotiate easily at the approaches. Some savings in cost was achieved but at about £2 million the cost was still above the desired level.

The arrangement tried out is shown in Figure 2.5.

The committee finally concluded that the balance cantilever bridge was the most appropriate solution, in view of the fact that in such a structure the primary load on the foundations is vertical in nature, and the only horizontal forces can be caused by wind on the superstructure or due to the effect of

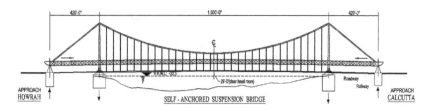

FIGURE 2.5 The self-anchored suspension bridge arrangement. (Ward & Bateson, 1947)

earthquake. The trial monolith foundations done for the King George's Dock in Kidderpore served as a guidance for assessment of the condition of the founding strata.

Balanced Cantilever Option with High Tensile Steel

Further study continued with the balanced cantilever arrangement that had been originally recommended. One important development with regard to structural steel materials used in bridges had been the introduction and acceptance of high tensile steel as against the conventional mild steel as per the then standard BIS 153 (Ward & Bateson, 1947).

While the maximum permissible tensile stress in mild steel was 9 tons per sq in, in high tensile steel the allowed stress limit is 12.65 tons per sq in, an increase of 40%. Use of this grade of steel, which was getting accepted and had been used successfully in the Chelsea Bridge reconstruction, proved to be a major help and a cost saver.

With the use of this high grade of steel, the tonnage of steelwork could be reduced in all components. This made a significant difference in forces to be generated during cantilever construction

During erection, the top chords, designed for compression in the case of the suspended span, had to take tension, and had to be designed for the same as well.

Furthermore, with the reduction in weight of steelwork, a substantial cost reduction could also be achieved in the foundations. With lighter individual members to be handled, there was a saving in the cost of cranes and other erection equipment.

Two more important changes were made to the 1922 design. The anchor span was reduced to 325 ft (99 m), even lower than the Mookerjee Committee recommendation of 350 ft (106.7 m). As the anchor span was not required to carry the deck weight, this proved to be of considerable advantage. The height of the towers was substantially reduced, by about 45 ft (13.7 m), and that also helped in reducing the cost (Figure 2.6).

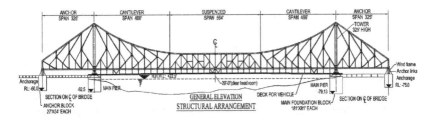

FIGURE 2.6   Bridge as was put to tender, and built

The new estimate of cost, based on the carriageway width of 60 ft (18.2 m), and pedestrian way of 15 ft (4.6 m), came to an eminently acceptable figure of £ 1.34 million, well within the budget. The government readily accepted this design, and the long gestation period of this bridge was finally over.

Meanwhile, the floating bridge continued to deteriorate, with an increase in volume of traffic and passage of heavier motorized vehicles. By 1933, it had been realized that the old bridge was no longer reliable. The possibility of war was looming larger every day. Calcutta was an important Allied military base, the new bridge assumed vital importance for movement of Indian resources – manpower and materials – which were very valuable to the Allies.

THE BRIDGE TAKES A FINAL SHAPE

When the news that the design had been finalized reached them, the Calcutta Tramway Company, foreseeing that money was to be made by trams using the bridge, came forward with the offer of meeting the extra cost of accommodating two tracks on the bridge, for up and down movement (Ward & Bateson, 1947).

A revised estimate was prepared with the carriageway width of 71 ft (21.6 m), making space for the tram tracks at the center of the carriageway. The revised estimate of £1.5 million was approved and accepted by the government and the stage was set for preparation of the final design, specifications, drawings and contract documents for inviting tenders for the bridge.

Invitation of Global Tenders

Rendel, Palmer and Tritton (RPT) the London-based consultants, appointed by the port authorities for the bridge project, started the preparation of the tender document.

For a major bridge of such dimensions to be built across a river having a heavy traffic of cargo vessels and passengers, in a heavily built-up urban environment with important structures all around the site, it was a major challenge to identify and select bidders capable of safe implementation of the project to a strict time schedule.

For ensuring a long safe life for the bridge with optimum use of materials, the consultants were given the responsibility for drawing up specification for materials, workmanship and loadings. Design and drawings, construction methodology and responsibility for supervision during execution, were in the scope of work of the consultants.

It was felt that the selected contractors must be fully involved in the project work at all stages for its success. It was therefore decided that, though the consultants would prepare the design ahead of inviting tenders and prepare the estimate and bill of quantities, the bidders would be given the option of submitting bids based on an alternative design to be prepared by themselves. This, it was felt, would prompt the bidders to make a thorough check of the design and construction methodology furnished with the invitation for bids, in their search for more economic alternatives for a winning bid.

The process of the actual construction activities began with the search for the main actor for this massive and challenging project, the construction agency, through a global tendering process.

REFERENCE

Ward, Arthur M. & Bateson, Ernest, *Paper No. 5601, Institution of Civil Engineers – The New Howrah Bridge, Calcutta: Design of the Structure, Foundations and Approaches*, 1947.

Preparation of Tender Document

3

LOOKOUT FOR BIDDERS

Major bridge contractors across the world were keenly attracted by the prospect of executing the proposed Howrah Bridge. The bridge was a technical challenge. It was a difficult bridge considering that it was over a very busy river in the middle of a great bustling city. The deep river caisson foundations in silt and clay were also a "dare." The road bridge in India itself may have been an attraction as only railways had built major bridges until then.

Apart from the technical challenges, in the impoverished and drained post-World War market, there were very few major bridge contracts going and bridge contractors were eyeing the developments of this project with avid interest.

There was yet another reason. The port authorities were giving the bidders the option of submitting alternative designs and schemes of construction, though the responsibility of preparing the complete design, working drawings, construction plan and bill of quantities were entrusted to the selected consultant, RPT.

Port authorities expected two advantages out of such a tender. One, a much-desired saving in cost may result if the competition produced a cheaper acceptable alternative. Two, in order to submit competitive offers, the bidders would go through the official design with a fine-tooth comb to find areas where they could reduce cost. This would incidentally provide a proof check of the design for the authorities.

Preparation of the tender documents was an arduous and elaborate task as it involved inviting an alternative design and scheme, giving the bidders all the obligatory conditions that had to be fulfilled, design loads and stress limits, material specifications and acceptable sources and other usual terms and conditions of a global tender for construction of a major bridge.

BID STIPULATIONS

The bid document was made ready in December 1934 and some of its stipulations were as follows:

The main span was to be 1500 ft (457 m).

The formation level of roadway was to be + 45.00 ft (13.7 m) with reference to adopted base level in order to maintain the required clearance for river crafts permitted to travel across that section during the highest floods (Ward & Bateson, 1947).

The alignments of the approaches were still not final, but the gradient of the approaches was not to be steeper than 1 in 35 to cater to the animal-driven carts, which transported most of the goods over the bridge.

The official design was for a bridge of the balanced cantilever type, with a suspended span between the two cantilevers. The carriage way or the deck width was to be of 71 ft (21.6 m), enough to accommodate eight lanes of traffic in those days including two tram tracks. Cantilevered pedestrian ways of 15 ft (4.6 m) were to be provided, one on either side. Alternative schemes offered by bidders had to adhere strictly to these dimensions and obligatory conditions.

There were to be no supports from the river for the erection of steelwork as no interference with the busy waterway was permitted.

DESIGN LOADS

The loads and load combinations were given in full detail.

All components of the bridge were to be designed for the following loads, applied over the entire bridge or part thereof, whichever produced the greater stresses.

Dead load
Live load (from vehicles on the deck and pedestrians on the foot way)
Impact load
Wind load
Seismic load

Erection and other construction loads during erection
Imposed loads from secondary effects (Ward & Bateson, 1947)

Magnitude of various loads

Live load: For the road deck, wheel loads from heavy vehicles were specified in addition to uniformly distributed equivalent loads for estimation of maximum stresses on structural components.

Special heavy boiler wagon loads, as were being used in London, were specified to be located on the outer lanes, considering requirements for maintenance operations.

Congestion loads of 100 lb. per sq ft (488 kg/meter square) were to be considered in such stretches that produced maximum stresses.

For the footway design, a uniformly distributed load of 100 lb. per sq ft (488 kg/meter square) was stipulated, for the design of the deck slab.

Impact Load: In principle, it was felt that impact would only affect the deck members and the hangers transferring the deck loads to the main trusses. The flexibility of the hanger arrangement would prevent any impact effect on the main truss, which was very stiff. A formula was given for the computation of the impact load to be applied only to deck members and hangers.

Wind loads: To be calculated using the data from the Weather Office on wind velocity and other parameters

Earthquake: Earthquake effects on each member were to be calculated by application of a horizontal load equivalent to 1/15th of the weight the member applied at its center of gravity.

Secondary effects that were to be considered:

➤ Stresses induced by distortion of the truss members and floor beams
➤ Stresses due to eccentricity in the connections
➤ Stresses due to unsymmetrical placement of live loads
➤ Stresses due to temperature variation – The range of temperature to be considered was 40 degrees Fahrenheit to 160 degrees Fahrenheit (4°C to 71°C)
➤ Bending stresses due to self-load of members applied at their center of gravity
➤ Stresses caused by friction at pins and the bearings.

As all loads do not occur simultaneously, combinations of various loads, logically likely to occur, were identified. Likely statistical frequency of load combinations was computed, and different limits of permissible stress values were given for use in the design.

MATERIAL SPECIFICATION

As mentioned earlier, it was found during earlier analysis of options, that use of high tensile steel would lead to a very significant reduction in size of members and consequently substantial savings in weight of steelwork, foundation and erection costs.

Note: Use of high tensile steel had been started in steel construction, with its use in the reconstruction of Chelsea Bridge, work on which was then (circa 1937) in progress.

As an authorized specification for the material was not yet available, specifications for chemical composition and physical properties of this grade of steel were developed in consultation with manufacturers and were shared with the intending manufacturers of this new grade of steel in Britain and India.

These specifications for high tensile steel were now included in the tender documents for the reference and knowledge of the bidders. The tables as included in the tender document are reproduced (Tables 3.1 and 3.2).

TABLE 3.1   High tensile steel

APPROVED CHEMICAL COMPOSITIONS							
	C	Mn	Cr	Cu	Si	P	S
Plates and sections. Plain manganese steel	per cent.	per cent.	per cent.	per cent.			
Manganese-chromium chromium steels — Home	0.30 (max.)	1.6 (max.)	–	0.30 to 0.60			
	0.30 (max.)	0.70 to 1.00	0.70 to 1.00	0.30 to 0.50	Not to exceed 0.20%	Not to exceed 0.5%	Not to exceed 0.15%
Manganese-chromium chromium steels — Indian	0.30 (max.)	0.80 to 1.20	0.40 to 0.70	0.30 to 0.60			
	0.23 to 0.28	1.00 to 1.30	0.50 to 0.62	0.30 to 0.60			
Rivets	0.18 to 0.22	0.60 to 0.75	–	0.30 to 0.60			

(Ward Arthur M & Bateson Earnest, 1947)

TABLE 3.2   High tensile steel

	YIELD STRESS: TONS PER SQ. INCH.	ULTIMATE TENSILE: TONS PER SQ INCH.	ELONGATION PER CENT.	
PHYSICAL PROPERTIES			LONGITUDINAL.	CROSSWISE.
Plates and sections. Specified	23.0 (min.)	37 to 43	18.0 (min.)	16.0 (min.)
Representative samples				
Plain manganese	24.5	37.6	25.0	–
(Home steel)	24.6	40.8	19.0	–
Manganese-chromium				
Home steel {	24.3	37.2	23.0	–
	24.6	41.0	19.0	–
Indian steel {	23.7	37.3	25.0	–
	25.6	39.8	22.0	–
Rivet bars.			On 8 dias.	On 4 dias.
Specified	–	30 to 35	22.0 (min.)	27.0 (min.)
Representative { (1)	–	31.9	27.8	–
sample bars { (2)	–	30.0	30.0	–
Driven rivets	Specified driven shear value 26.0 tons per sq. in. (min.).			
Rivets from { (1)	Driven shear values 26.77 to 27.4 tons per sq. inch.			
above bars { (2)	Driven shear values 25.47 to 27.7 tons per sq. inch.			

(Ward Arthur M & Bateson Earnest, 1947)

High tensile steel was specified for all highly stressed members and for all main joint rivets, and steel to BS 153 (popularly known as mild steel) was specified for deck members subject to bending stresses, and secondary and lightly loaded structural members.

The maximum permitted working stress values for tension, compression, bending etc. for different grades of steel were clearly given in the tender documents as reproduced (Tables 3.1 and 3.2).

Reduced stress values for extra-long rivets, and for rivets driven at site (i.e. other than those done in controlled conditions in the shops) were stipulated in the documents.

Rivet heads for rivets of the two grades of steel allowed for use were required to be made with a different shape to avoid an unintended mix up and likely danger of weak joints (Ward & Bateson, 1947).

DESIGN OF COMPRESSION JOINTS

Certain design guidelines were provided to the intending bidders for their use if found to their advantage.

Foremost amongst them was that for the design of compression joints – the load transfer through butting of (machined) joints could be taken as 50% and the balance 50% through the rivets in the joints. However, this would not apply if (a) where sections needed to be augmented due to high secondary bending stresses and (b) machining facilities were not adequate to achieve the standards required. In these cases, the joints were to be designed for full transfer of forces through rivets (Ward & Bateson, 1947).

TENDER DOCUMENTS FOR FAIR BIDDING

The tender document was elaborate and detailed. It was intended for providing a fair and level playing field to the bidder. This was evident as the basis of the official design had the same specifications for design, materials etc. as given in the tender documents for bidders to adopt.

All the bidders were provided with the design specifications and the conditions of contract in draft form, giving them the freedom to submit their suggestions for revisions.

The bidders were invited to visit the site to know and familiarize themselves with the site conditions and limitations, gather required field data and have discussions with representatives of the owner, the commissioners of the port, and their consultants. Adequate time was provided for the submission of bids allowing for ship journeys to and fro, for site visits (Ward & Bateson, 1947).

The final tender document was released in December 1934 and the submission of the bid at Calcutta, or at London, was fixed as March 1935 but later extended at the request of the bidders to 30 April 1935.

Four bidders, two from Britain, one from Germany and one from Calcutta, submitted their bids, and were duly accepted.

SOCIO-POLITICAL DEVELOPMENTS

Some new political and commercial developments impacted the review of the bridge tenders. While the pre-construction and tender activities for the bridge were in full swing, new and unprecedented developments were taking place, even amongst steel fabrication firms and steel manufacturing units, in India.

The movement for independence had been rapidly gaining strength in India in the thirties. The Government of India Act would be passed in 1935 giving a good measure of self-government to India.

This being the political situation, the British-owned commercial and industrial organizations in India wanted to avail themselves of the opportunity to secure greater business for themselves. The general consensus was growing that there was a real hazard of giving all the business to Britain-based firms.

The largest steel manufacturing unit in India, Tata Steel, located at Jamshedpur, close to Calcutta, made a very strong representation to supply steel for the bridge. They created facilities for manufacturing the required high grade of structural steel. They were able to achieve the quality required and with that they were able to convince the client and the consultant that they would be able to supply the quality and quantity of steel required. Tables 3.1 and 3.2 indicate the results of efforts to develop high tensile steel by steel manufacturers in Britain and India. What is marked as Indian steel in the tables is the result of tests done in Tata unit and one can see the results match or are superior to steel from UK units. Ultimately, Tata Steel supplied most of the steel for the bridge.

Three major fabrication units of India, Braithwaite & Co., Burn & Co. and Jessop & Co., all having high quality fabrication facilities in and around Calcutta, decided to join forces and make a strong pitch for the tender for the bridge to be built in their back yard. They floated a new company named The Braithwaite, Burn & Jessop Construction Co. Ltd (BBJ) for participating in this project. The new company secured orders for a number of bridges by way of building up credentials. One of the bridges that was fabricated and erected by the new firm was a railway bridge across River Meghna in what is now Bangladesh.

For producing alternative designs, BBJ secured the services of noted bridge designer Mr. Freeman (later Sir Ralph Freeman, the senior partner of the iconic bridge design firm Freeman Fox and Partners). Mr. Freeman produced various alternative designs, which allowed BBJ to produce the lowest bid.

COMPARISON OF TENDERS

Bids were submitted, by the British and the German firms, based on the official design. BBJ, however, submitted one bid according to the official design and six more bids with alternative designs.

The financial bids were very close, giving the authorities greater confidence in the official scheme that had been put to tender.

While the bid value of the four tenders based on official designs varied from Indian Rs 20.9 million to 23.3 million, the bids on the alternative designs varied from Rs 18.2 to 22.8 million (Ward & Bateson, 1947).

The German tender, based on the official scheme, was the lowest and, in normal circumstances, this bid would have been the automatic winner. However, as in the case of the tender for the pontoon bridge, war clouds were again gathering, and political considerations dictated the scrapping of the lowest bid. As feared, the deadly war did break out in September 1939 and the bridge had to be constructed and completed under the shadow of the war.

BBJ had submitted three alternative bids based on the official design of superstructure but modified the foundation design and one of them was the lowest bid. In one case the official design of the foundations was followed, but changes were made in the construction methodology and concrete materials. The other two bids did not follow the official scheme of multi-chamber caissons but provided multiple caissons.

The remaining three alternatives used changes in superstructure design as well as modifications in the foundation arrangement. The modified superstructure design was shorter in length by 46 ft (14 m) than the specified 1500 ft (457 m) but offered only a limited benefit in price. The modification in the foundation design offered significant savings in cost but the consultants did not find them acceptable. The modified designs proposed in the superstructure were rather futuristic for those days, including an orthotropic design for deck grillage and use of cables in place of steel structure for part of the top chord of the cantilever section – a hybrid solution. These, however, were not found to be in tune with the accepted practice of the time and were at variance with the design specifications issued and were therefore rejected.

BBJ was not satisfied with the analysis of their bids, as prepared by RPT, and approached the owners of Calcutta Port for review. As directed by port authorities, RPT reviewed the detailed analysis made by BBJ while preparing the tender but perpended with a very detailed report in January 1936 rejecting BBJ's arguments in defense of their bid (Report submitted by M/s. Rendel, Palmer & Tritton, 1936).

The port authorities agreed with RPT and from amongst the remaining bids based on official design, the Cleveland Bridge and Engineering Co. Ltd of Britain was selected, being the lowest amongst the two bids found acceptable.

The story of selection of the implementing agency did not end there. There was a very strong move for involving BBJ in the bridge project, by the local industry circle, and the Chambers of Commerce took up the issue strongly on behalf of the local firm. Ultimately, a compromise solution was found, by making BBJ the nominated subcontractor for the fabrication of all the steelwork, to be carried out at the works of the three constituent firms, Braithwaites, Burns and Jessops. The responsibility for overall planning and implementation remained with Cleveland Bridge Co.

This decision was to have a great influence on the future of bridge construction industry in India. There was a very effective transfer of technology to the Indian industry for steel fabrication, with the local industry vying to achieve the exacting quality demands in the project. Cleveland Bridge Co. was relieved of the worry of transporting the fabricated pieces from Britain across the seas (running the gauntlet of German U-boats) and the problems of assembly of the joints, ahead of erection.

At last, the actual construction was ready to take off, and the project, which had been very long in planning, was to take shape.

REFERENCES

Rendel, Palmer & Tritton, *Report on Consulting Engineers on The Counter-Criticisms of The Braithwaite, Burn & Jessop Construction Company Limited, New Howrah Bridge,* January 1936.

Ward, Arthur M. & Bateson, Ernest, *Paper No. 5601, Institution of Civil Engineers – The New Howrah Bridge, Calcutta: Design of the Structure, Foundations and Approaches,* 1947.

Design of Superstructure

4

CHALLENGES

Designing a 1500 ft (457 M) span superstructure and as a balanced cantilever was a big challenge in the 1930s. The other two long span balanced cantilever bridges that were predecessors to Howrah Bridge were the Forth Bridge in Scotland and the Quebec Bridge in Canada. Both were located in rolling terrain, at locations with no habitation or built environment in the vicinity, and the railway/roadway was taken through the mid-level of the truss structure, along the axis of the bridge.

Howrah Bridge, on the other hand, was in flat country, and the roadway formation level had to be maintained as low as possible by maintaining the required clearance from the highest flood level in the tidal river. The deck structure needed to have a minimum possible depth in order to reduce the formation level in order to minimize the length of the approaches. Further, in this fast-growing city, the approaches had to be maintained close to the river bank without affecting the various buildings. The gradients of the approaches had to be not more than 1 in 36, for easy and safe negotiation by the bullock carts that carried most of the goods across the river.

These problems were addressed by adopting a grid-form deck suspended from the main truss by vertical hangars that transferred all the deck loads to the main truss. The traffic from the bridge was taken down to the bank level along the approach road that ran almost parallel to the river on Calcutta bank, and on the Howrah bank through a loop around the anchorage structure, the up and down roadways being separate.

The approach roads were made independent of the bridge structure, by taking the traffic flow through portals on the tower, then through a near right angle turn, to pass under the bottom chord of the truss in the anchor span stretch.

While the rear end of the approaches, i.e., up to about 6 ft (1.8 m) height, was laid on consolidated fills, the front end of the approaches was built as a flexible structure, connected by pins on to the bottom strip footings, laid on a well-consolidated base course. The flexibility was needed to take care of possible settlement of the back-filled area, around the tower foundations. It would also allow for seismic movements (Ward & Bateson, 1947).

Stability against such induced movements was achieved by connecting the strip foundations with reinforced concrete struts and ties. Portals were provided with hinges at base and at the junction of the columns and beams that supported the deck.

The towers, 280 ft (85.3 m) high, connect the anchor truss and the cantilever truss. The suspended span, 564 ft (172 m) long, is freely supported on the cantilever arms, through pinned connections.

The truss arrangement had to be statically determinate to minimize the secondary stresses.

It was necessary to limit the weight of steelwork from several aspects such as cost, excessive moments due to the long cantilever, very deep tower foundations and so on. Fortunately, a solution was readily at hand through the use of high tensile steel, which was just then being tried out successfully in Chelsea Bridge, underway in UK. It was decided to use this higher grade of steel. Steelwork tonnages were reduced and there were consequent savings all-round in the tonnage of steel requirement, which contributed to a significant overall reduction of the cost of the bridge (Ward & Bateson, 1947).

For reducing the secondary stresses in the primary members of the truss, pre-stressing of the steel members was introduced, for the first time in a major bridge structure. The length of the components of the superstructure was so fabricated that under the dead load and the full live load, the superstructure will assume its true geometric shape, eliminating the secondary stresses under the maximum loading condition. This innovation permitted use of lower sections of members, leading to further savings (Figure 4.1).

The erection scheme of the superstructure, across the busy tidal river, had to be conceptualized at the design stage itself, in order that all components could withstand erection stage stresses. It was necessary to work out the likely sag at the end of cantilever arm and provide suitable closure arrangements. It was reckoned that cantilever erection was the only feasible plan of work, and the design was developed accordingly. Bidders were given the option to erect the central suspended span with the help of floating cranes but the final choice

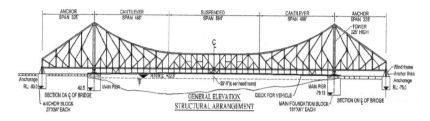

FIGURE 4.1   Elevation arrangement of the bridge

of the contractors was to erect the suspended spans also by cantilevering from the arms at either end, as had been envisaged by the consultant.

STRUCTURAL ARRANGEMENT

Truss Geometry

The structural arrangement adopted was a perfect example of a balanced cantilever suspension bridge. The anchor spans at either end, with spans of 325 ft (99 m), between the tower and the anchor block were required to support the 1500 ft (457 m) central span that comprised two cantilever sections of 468 ft (142.6 m) at either end, which in turn supported the central suspended span of 564 ft (172 m). Refer to Figure 4.4 for the structural arrangement.

The weight of the concrete anchor foundations provided the stability against overturning. Considerable study and analysis preceded the finalization of the lengths of different components of the trusses, and the height of the towers, to arrive at the most economic weight of the bridge structure (Figure 4.1).

K-trusses were finalized for the anchor span as well as the cantilever arm. It was divided into subpanels to reduce the slenderness ratio of long

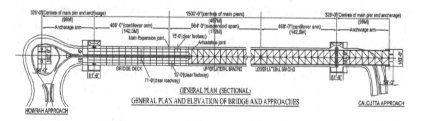

FIGURE 4.2   Plan arrangement of the bridge and approaches

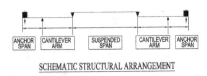

SCHEMATIC STRUCTURAL ARRANGEMENT

FIGURE 4.3   Schematic representation

compression members. Panel lengths of the trusses were decided on the basis of economy of the deck structure. K-trusses were determined by the proportion of the height of the towers, vis-à-vis the panel length. The panels adjacent to the towers, were made longer (42.75 ft/13 m near the towers, compared to the 35.25 ft/10.75 m panels at the central section), for a proper acceptable slope of the diagonals. For the suspended span of much shorter height, sub-divided N panels were used.

DECK ARRANGEMENT

The deck width of 71 ft (21.6 m) between curbs allowed for 8 lanes of traffic including 2 tram tracks. The side lanes near the truss were paved with cast iron strips, to prevent damage from the steel tires of bullock carts. Foot paths of 15 ft (4.5 m) on either side were cantilevered from the truss girders. The large width of the footpaths was to cater for the anticipated future volume of pedestrians arriving at the city from the suburbs by train for their livelihood. The total width of the deck structure was 108.75 ft (33 m) between centers of fascia girders at the outer edges of the pedestrian decks.

For the deck, a system of cross girders, stringers, cross beams and pressed steel troughs were used. Cross girders spanned the deck between verticals hung from the truss at panel points. Stringers spanned longitudinally between cross girders. Cross beams were used between stringers. Pressed steel troughs were laid along the length of the bridge on cross beams. Three inch (7.5 cm) thick reinforced concrete slab, laid on the troughs formed the base course, over which a concrete wearing coat, 2 inches (5 cm) thick, was provided on top of the deck slab, secured by steel hoops, which were welded in-situ with the trough and the reinforcements of the deck slab. This unconventional arrangement of wearing coat proved very effective and durable for withstanding the pounding of very heavy traffic over years (Ward & Bateson, 1947).

The clearance from the top of the finished deck to the bottom of lateral steel bracings between trusses was kept at a minimum of 19 ft (5.8 m), which was considered adequate for all vehicles likely to ply on the deck in future.

The center-to-center distance between the hangers on either side (and also the trusses) is 76 ft (23.16 m). Pin joints were used to connect hangers to the trusses, ensuring flexibility of the deck and limited transfer of forces from the rigid deck to the truss girders.

Continuity in the 1500 ft (457 m) long and 71 ft (21.6 m) wide deck would have created bending stresses in the verticals and secondary stresses in the superstructure truss members, apart from the temperature effects on the deck steelwork. Therefore, expansion joints were provided dividing the deck system into three distinct sections a) at the towers b) at the junctions between cantilever arms and the suspended span. These sections were again subdivided further, in four laterally articulated sections in the cantilever arms and in three sections for the suspended span.

The main expansion joints were provided with cast steel, toothed-intertwining sections, covered by moveable checkered plate on top.

TOWERS

The towers are the most visible and the most critical load transferring item in the bridge structure.

Because of the balanced cantilever arrangement, the towers carry more load than the total weight of the structure and the live loads imposed on the carriageways. The towers provide the dividing line between the anchor arm and the cantilever arm.

The towers at each end are composed of two posts, 279.75 ft (85.27 m) high. K-form lateral bracings are provided between the two ports, above the top of the cross beam at the bottom chord level, below which the traffic passes. The posts appear vertical when viewed from the river, but tapering out towards bottom, as viewed from the deck (Ward & Bateson, 1947).

The posts, at the saddle level, connect with the top chords and diagonals through pins (Figure 4.4). At the bottom chord level, the truss rests on the main bearing, which is supported on a bracket, cantilevering out inwards from the post. The posts are 76 ft (23.2 m) cross centers at top but splay out to 95.67 ft (29.2 m) at the bottom, to accommodate the lower chords passing through them. The bearings are connected to the bottom chord and the diagonals meeting at the bearing point, through pins, allowing flexural rotation at the support (Figure 4.7). The bearings rest on rolled manganese-bronze bearing plate fixed on a 4 inch (10 cm) thick forged steel capping slab that is supported on the cantilevered bracket, allowing sliding movement.

The posts are rectangular in cross section, with three vertical web plates and one transverse vertical diaphragm, dividing the cross section in eight

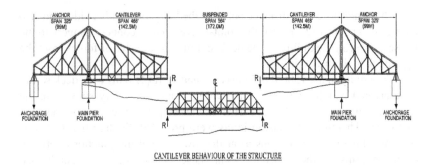

CANTILEVER BEHAVIOUR OF THE STRUCTURE

FIGURE 4.4   Schematic representation of balanced cantilever arrangement

chambers. With a constant side elevation width of 11.5 ft (3.5 m) above the bearing level, the posts taper on the transverse side from 8.5 ft (2.6 m) at the bearing level to 4.5 ft (1.4 m) at the top. There are horizontal diaphragms at intervals of 9 ft (2.7 m). The section of the posts below the bearing level flare out in both directions to have a base dimension of 24 ft (7.32 m) square and rest on a series of grillage girders, which transfer the load on to the concrete pier.

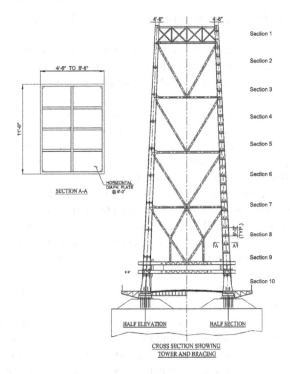

CROSS SECTION SHOWING
TOWER AND BRACING

FIGURE 4.5   General view of tower

ANCHOR STRUCTURE

The anchor structures at the ends provide the vertical stability to the balanced cantilever arrangement and transfer the lateral loads to the foundations.

The end post of the truss, designed to take the vertical uplift at the ends, is connected through a pin to a vertical link member, which is anchored inside the concrete block, by pin connection to a set of three 9 ft (2.7 m) deep girders. These girders are held in place by bearing on grillage beams, embedded at both ends in the walls of the concrete monolith of size of 27 ft (8.2 m) × 54 ft (16.5 m), sunk to a depth of 87 ft (26.5 m) for the Howrah side and 104 ft (31.6 m) for the Calcutta side.

A pair of braced A-frames connected to the lower lateral cross strut of the anchor truss, through pins, have been provided to take the lateral wind reaction from the superstructure, without interfering with the longitudinal and vertical movements, resulting from the variations in temperature and deformations due to lateral loading.

Provisions for adjustments had been made inside the anchor monolith by providing arrangements for jacking operation, such that corrective measures can be taken in case of settlement of foundations of the main towers.

FLEXIBILITY OF STRUCTURE ALONG BOTH AXIS

Great care had been taken to ensure that the behavior of the structure conformed to the design assumptions. Analyzed as a statically determinate structure, provisions had been made to ensure such behavior by using pinned connections at important joints. There are 13 pin joints in each half of the truss, as shown in Figure 4.8. They include the connection of the truss with the tower posts and the connection between the cantilever arm and the suspended span. The towers are sufficiently flexible to be part of the truss (Figures 4.6 & 4.7).

The deck was hung from the truss through pin connections, ensuring freedom for the trusses from the laterally rigid deck structure.

The lateral bracing system, in the lower chord level, was connected to the cross strut in the tower, through vertical pins at the central axis, thereby freeing the lateral system from the effect of truss chord deformations. This also ensured the suspended truss stayed stable along the longitudinal axis. The A frames at the anchor end frame transfer lateral loads through pins at the central axis and ensure that assumptions made in the theoretical analysis remained valid.

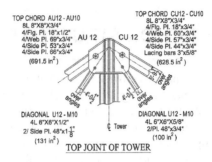

FIGURE 4.6   Pin joints at top of tower (Ward & Bateson, 1947)

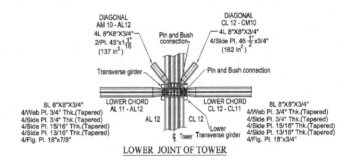

FIGURE 4.7   Lower chord joint at tower with pins (Ward & Bateson, 1947)

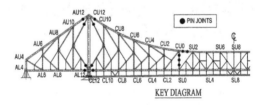

FIGURE 4.8   Key diagram showing pin joints

The top chords of the anchor arm and the top chords of the cantilever arm remained permanently in tension and therefore no bracing system was provided along the top chord. Complete sway frames were provided at all main and sub verticals. A robust bottom lateral system was provided to ensure transfer of lateral forces to the anchor frames at ends. Lateral wind forces on the suspended span were transferred to the cantilever arms through articulated joints at the central axis and at lower chord level. For preventing longitudinal oscillation of the suspended span, friction grips were provided at the junction with cantilever arm.

FIGURE 4.9   Bracing system adopted

In the suspended span, bracing systems were provided along the top and bottom chords, and sway frames were provided at all verticals (Figure 4.9).

DESIGN FOR EARTHQUAKE FORCES

In the 1930s, knowledge about the behavior of structures during an earthquake was very limited. However, in the present case, the design specification provided for consideration of a lateral force, applied at the center of gravity of every member, equivalent to G/15.

The design was not governed by the earthquake forces as other load conditions produced worse effects. The design was created keeping the effect of earthquakes in mind (Ward & Bateson, 1947).

PRESTRESSING OF STEEL MEMBERS

Considering the very large truss structure for the bridge and the effects secondary stresses can cause, it was decided to introduce pre-stressing of truss components. All the members were manufactured by modifying their length according to the strain produced by dead load plus full live load conditions.

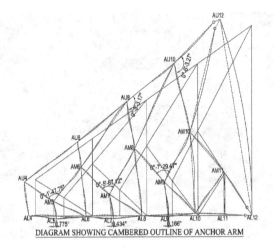

DIAGRAM SHOWING CAMBERED OUTLINE OF ANCHOR ARM

FIGURE 4.10   Diagram showing the cambered outline of the anchor arm (Ward & Bateson, 1947)

The joints were laid out and fabricated in keeping with the geometric outline and the intersection angles were laid true to the truss geometry. This necessitated application of certain additional force to introduce curvature in the member, while completing erection of a triangle with fabricated members. The direction of curvature and angular rotation of joints were calculated along with the force required to create these deformations. Special attachments like pull-lifts and jacks were arranged for use during the erection process (Figure 4.10).

The superstructure was cambered, using a basic analytical tool like the Willot Mohr diagram for dead load plus full live load condition such that in a loading condition the superstructure attained perfect geometric configuration, ensuring zero secondary stress.

For other loading combinations, secondary stresses were calculated and added to the primary stresses. However, they did not prove to be critical for design.

No difficulty was encountered in fitting in the members during the erection process and the truss developed a theoretical geometric shape under a full live load.

DESIGNING FOR ERECTION CONDITION

During the design stage itself, attention was given to the method of erection of the bridge. Considering the constraints of the site and the need for keeping the waterway open at all times, the cantilever method of erection was found to be the ideal solution. While designing the members of the truss, both the service load condition and erection condition forces were calculated, and members designed for the critical force. All aspects of erection condition, including the positioning of crane and its imposed load, as well as impact effect, were considered for different stages. For the erection of the suspended span, apart from the cantilever erection with a crane on the top chord, the possibility of lifting the span using floating cranes was kept in mind (Ward & Bateson, 1947).

Special strengthening of members and designing with modified sections was performed as necessary. The hinge joint at the junction of cantilever arm and the suspended span needed special care during the erection stage.

Bidders were issued the erection scheme as considered during design for their examination, and they were given the freedom to develop alternative methods of erection. To the credit of the design consultants, no bidder came out with better schemes as an alternative.

It is a tribute to the designers that, with all these special requirements, and some of them unique for the period, the design and specifications were completed in less than a year with no help from advanced design tools as available today.

REFERENCES

Ward, Arthur M. & Bateson, Ernest, *Paper No. 5601, Institution of Civil Engineers – The New Howrah Bridge, Calcutta: Design of the Structure, Foundations and Approaches*, 1947.

Design of Foundations

5

CHALLENGES

In the 1930s building a large span bridge in a virgin area where no heavy structure of this kind had been built earlier was in itself a major challenge. The only important engineering project in the vicinity that needed heavy underground work was the Calcutta Port, and some information on the underlying soil strata was available from that project. It was known that up to large depths the available strata were silt, clay and sand, and no firm foundation strata would be met.

The science of soil mechanics was still in its formative stage and methods of investigation had not been developed nor were tools like the Standard Penetration Test or similar yet available. Investigation procedures had not been standardized. Designers had to develop their own methods for ascertaining the safe carrying capacity of soil by site investigation. In the present case the loads for which foundations were to be designed were enormous by any standard.

The loads on the foundations were primarily of vertical nature as the balanced cantilever design had been adopted for the bridge. The imposed load on the tower foundations were considerable, at around 30,000 tons, and the vertical uplift at the anchorages was about 13,500 tons. No indication of safe bearing pressure of the soil at different depths was available to the designers.

The river was in the flat deltaic region and in the meandering stage. It could be expected that the soil strata in the proximity of the banks would be different from the main channel with river deposits displacing the usual silty clay and clay strata with fine sand.

At the selected bridge site, the substrata on the Calcutta end were considerably at variance with those at the Howrah end. On the Howrah side, the deposits were undisturbed and in virgin condition consisting generally of soft clay and silt, until sand was met 60 ft below ground. After penetrating 19 ft of water surcharged sand layer, a stiff clay layer was reached. On the Calcutta bank, the stiff clay layer was reached at 97 ft (29.6 m) below ground, and the layers above were soft silt and silver sand, apparently deposited by river flow in the not so distant past.

INVESTIGATION

After perusal of the soil data available from the Calcutta Port construction work and after studying the data obtained from the borings carried out around the foundation locations, the designers decided that the foundations would have to be taken down to the stiff grey clay layer at depths of 60 ft (18.3 m) or more. But as there were no theoretical guidelines available for arriving at the safe bearing capacity of clay layer, recourse was taken to field tests (Figure 5.1).

Two types of tests were carried out in sequence. Initially, a steel screw pile with a 4.5 ft (1.4 m) diameter cutter at the bottom was taken down into the clay layer and then loaded to an extent of 4.5 tons/sq ft (48.4t/meter sq), which was more than the expected bearing pressure. This test was satisfactory with no adverse indications. A second test was then taken up.

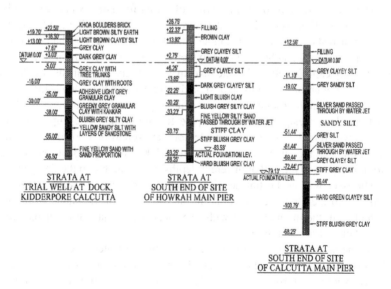

FIGURE 5.1   Soil Strata at Trial Boring and at Pier Foundations (Ward & Bateson, 1947)

Cast-iron cylinders of 6 ft (1.8 m) diameter, with flanges turned inside, were sunk on both banks near the tower location, well inside the stiff clay layer. On the Howrah side, the cylinder was taken down to 92 ft (28 m) below ground level. The stiff clay layer was met at about 80 ft (24.3 m) below ground and sinking of the cylinder was continued for a further 12 ft (3.66 m) considering the need for proper anchoring of the foundations. The hole created was dewatered and the soil at the bottom was physically inspected by persons going down to the bottom. Thereafter, the cylindrical hole for its entire depth was filled with concrete and so loaded as to create a base pressure equivalent to the planned level of about 20 tons/sq ft (215.3t/m²) and no settlement was observed. Loading was further continued in order to ascertain the safe capacity at that level, and after reaching a loading of 21.5 t/sq ft (231.4 t/m²), the clay showed settlement (Howarth & Shirley-Smith, 1947).

On the Calcutta side, the same procedure was adopted. But physical inspection of the soil could not be carried out as water gushed in from the water bearing sand at about 114 ft (34.7 m). However, the load testing could be done by filling the cylinder with concrete fully and loading up to 17 t/sq ft, (183 t/m²). There was no settlement whatsoever. Settlement was observed only when the loading reached the level of 25 t/sq ft (269.1 t/m²) at the base.

These two tests gave a good indication of the depths at which safe and reliable bearing capacity was available. It may be noted that the imposed load on the foundations both from the dead and live loads was about 1.8 t/sq ft (19.4 t/m²) and the addition from the gross load of the foundation structure was about 4 t/sq ft (43.1 t/m²) on the Howrah side and 4.7 t/sq ft (50.6 t/m²) on the Calcutta side. It was reckoned that the weight of the foundation structure, including water filling inside the concrete box, was more or less the same as the weight of the soil strata displaced. The added loads on the foundations were from the gross weight of the structure and the live load or the load imposed from traffic.

DESIGN OF THE TOWER FOUNDATIONS

Considering the loads that were to be carried, it was concluded that it would not be possible to rest the foundations on piles. During those times, the maximum diameter of driven in-situ piles was 24 inch (60 cm) and the installation of bored piles was ruled out in view of soil profile encountered.

That left large size caissons that could be sunk by dredging of soil from inside, as the only practical solution, for foundations. It was considered essential that the base of the dredged wells be inspected for the consistency and quality of soil strata before the founding level was decided.

It was also necessary that tilt and shift of the large caissons, when taking them to their founding level, should be restricted to the absolute minimum possible. The founding level was expected to be at least 90 ft (27.4 m) below ground. A multi-chamber cellular monolith construction was therefore decided on for the foundations.

The size of the foundations was designed to ensure that there was a sufficient factor of safety, in case the available bearing capacity was less, and also to limit settlement in the foundation block.

After deliberations, a single monolith of a size of 181.5 ft (55.32 m) by 81.5 ft (24.8 m) was decided upon with 21 chambers (seven along the river and three in the transverse direction) each 20.5 ft (6.3 m) square. Later, smooth progress of construction at site justified the adoption of this multi-chamber arrangement that allowed systematic progress of dredging without any significant tilt or shift of the caisson, and which had adequate weight to overcome the skin friction on walls and the resistance to the cutting edges under the peripheral and inner partitions walls (Ward & Bateson, 1947) (Figure 5.2).

The monolith walls were designed to resist the lateral pressure from soil and water and the stresses imposed during sinking, for both conditions of water inside the well and without water. The walls were provided with a steel curb and cutting edge at the bottom for the caisson to cut through the soil as dredging is done from inside the chambers. The external walls were 5.84 ft (1.78 m) thick and the internal partition walls 5.3 ft (1.6 m). The weight of the caisson of these dimensions was adequate to overcome the resistance provided by soil during sinking operations (Figure 5.3).

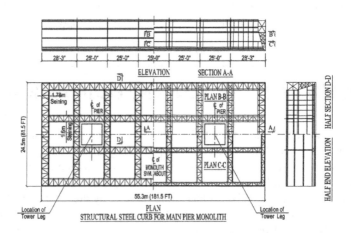

FIGURE 5.2   Layout of main pier monolith curb (Ward & Bateson, 1947)

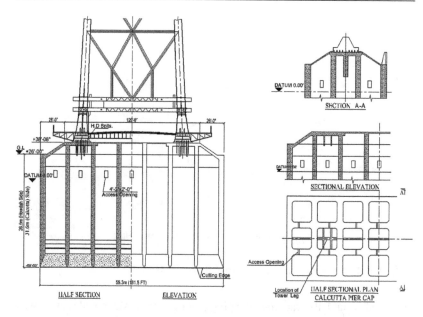

FIGURE 5.3   Arrangement of pier foundations (Ward & Bateson, 1947)

The curb was kept adequately wide to support the shuttering plates for casting of the first lift of the walls. Provision was made for moving the shutters upwards for casting of subsequent lifts of the walls.

Because of the large size of the monoliths, sinking with the help of kentledges was considered impracticable. It was therefore decided to fit the curbs with steel nozzles, so that pressure jetting could be applied, if found necessary. These were, however, not needed to be put into use during construction.

The curbs had to take the stresses that could be imposed due to unequal resistance from soil over the periphery of the caisson, and also from unknown obstructions that could lie underneath, like sunken crafts, and other debris (Figure 5.4).

The curbs were designed as continuous girders, as composite structures with the infill of concrete, for the sinking operations. Care was taken to make the ground surface level and as far as possible underground obstructions were removed after prior investigation. Once the caisson sank sufficiently, the stresses reduced and the movement became uniform in nature (Ward & Bateson, 1947).

Fearing the possibility of "sand boiling," when the foundation block moved through water surcharged fine sand layers, provision was made for fixing airtight covers on top, to convert each chamber into a compressed air working area, whenever the need arose. Application of compressed air, it was

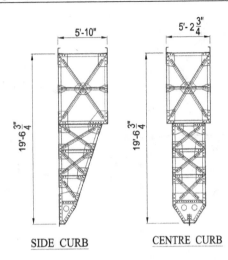

SIDE CURB   CENTRE CURB

FIGURE 5.4   Pier caisson curbs (Ward & Bateson, 1947)

envisaged, would also facilitate the plugging operation. Machined seating was fitted on the inner edge of each dredging shaft, to ensure that covers were air-tight and for proper functioning of the compressed air operation.

This was the first use of compressed air aided sinking of caissons in India, and one of the early ones across the world. Adequate preparatory measures were therefore planned ahead. The use of compressed air in the dredge holes allowed inspection, at any time, of the inside of the monolith and the soil strata. Gangways were provided along the internal wall surface to allow movement of inspecting personnel.

Access openings of size 4′2″ × 2′0″ were provided on the walls of the caisson to facilitate movement between the chambers by workmen.

All these preparatory measures were rewarded with smooth unhindered sinking of the monoliths.

DESIGN OF ANCHORAGES

The anchorages were to be subjected to downward loads only during the initial construction stage of the anchor span which taken up first. They were subject to increasing uplift force as the cantilevering of the superstructure progressed.

The basic design requirement was to provide a factor of safety of 1.5 over the maximum uplift forces that could be imposed at any stage of construction or service condition.

The loads, downward or uplift, were to be applied through the axis of main trusses, spaced 76 ft (23.2 m) apart. The loads to be resisted permitted the two anchorages to be separate, not combined as in the case of the main piers (Ward & Bateson, 1947).

Two separate monolith blocks, each 27 ft (8.2 m) by 54 ft (16.4 m), with two separate dredge holes were decided on. The blocks were connected with each other at the upper level.

The design considerations for the reinforced concrete walls were guided by the lateral loads caused by earth pressure and the stresses imposed during sinking operation. Provision for sinking under compressed air was made as was done for the tower monoliths (Figure 5.5).

Anchorage girders were provided for holding the uplift forces from the superstructure and the transfer of the forces to the monolith block was made through steel grillage beams embedded in concrete.

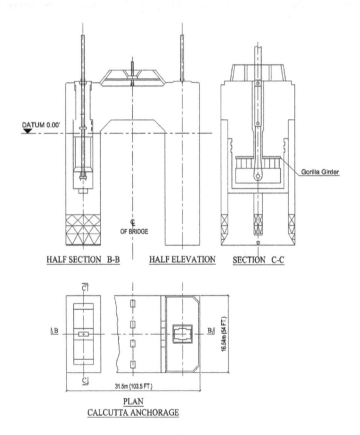

FIGURE 5.5 Arrangement for anchor block

Provision was made for future adjustments in levels of the tension links inside the anchorage block, to cater for likely settlements of towers or unplanned deflections of the cantilever arm of superstructure. Access for inspection, during construction and during the service life, was planned in the anchor block. Sumps were provided for draining seepage water to prevent corrosion damage to the steel items inside.

REFERENCES

Howarth, George Eric & Shirley-Smith, Hubert, *Paper No. 5612, Institution of Civil Engineers – The New Howrah Bridge, Calcutta: Construction*, 1947.

Ward, Arthur M. & Bateson, Ernest, *Paper No. 5601, Institution of Civil Engineers – The New Howrah Bridge, Calcutta: Design of the Structure, Foundations and Approaches*, 1947.

Construction of Foundations

6

Once design and drawings were final and the successful bidder was at the site, the construction of the bridge project commenced. Construction started with the foundations for the towers and the anchor blocks at both ends of the bridge. All the foundation structures for this giant river bridge were located on land and that was a great advantage for the builders. However, there were other challenges to meet.

CHALLENGES

For the selected construction agency from Britain, the soil stratification in the deltaic region was a new exposure. The river in its lower reaches, before confluence with the sea, about 100 km away, flows through silt and sand (refer to Figure 5.5 of Chapter 5). In between silt and sand layers, there were layers of clay of varying depth and consistency.

The ground water level was at the surface during the rainy months. The river reach was subjected to tidal flow four times a day, with the river water level varying by up to 20 ft (6.1 m) maximum towards the end of monsoon months.

The foundation work involved sinking large rectangular caissons to a depth of about 100 ft (30.5 m) through varying strata. There was no past construction experience of a similar activity in nearby areas. But caissons had been sunk for the berthing walls for Kidderpore dock of Calcutta Port, for the rail cum road bridge at Bally, about 10 km upstream, and for the rail bridge over Hooghly, about 40 km upstream. Problems had been faced in sinking caissons for both these bridges due to tidal movements of the river. In the present case, though the foundation locations were just outside the river cross section, they were affected by tidal flow. This had been noted and the planning had considered possible problems on this account.

The soft soils and varying strata through which the caissons had to be sunk made it likely that tilts and shifts of the caisson could occur and correcting them for a mammoth caisson would be a very difficult proposition.

The smooth sinking of caissons is hampered by obstruction due to junk materials, from sunken rivercrafts or materials accumulated inside the river. No technology was available to detect such obstructions, ahead of the sinking operation. Keeping a very close watch on the sinking operation was the only way of preventing, or at worst, limiting to the very minimum, tilts and shifts.

Fine sand strata surcharged with ground water induces sand boiling, and for effective sealing of the bottom of the caissons with impermeable mass concrete, it was essential to make the inside dry. Arrangements had to be made for such an eventuality prior to commencement of work. A pneumatic sinking process was envisaged and provided for, in anticipation that such a situation might arise. Though the technology was in its early stage of development, nevertheless it was decided to use it to ensure smooth and controlled sinking of the caissons.

It was decided that for a uniform sinking operation, removal of earth from the multiple chambers would be done by manual grabbing with workmen sent with digging tools, up to the maximum depth possible, and then only mechanical grabbing with the help of cranes undertaken. This proved very effective for ensuring vertical sinking of the caissons (Howarth & Shirley-Smith, 1947).

CONSTRUCTIONS AT SITE: PREPARATORY WORKS

Contractors mobilized at the site in October 1936. It was planned that work would simultaneously proceed on both banks and arrangements for materials, equipment and workmen were made accordingly.

All materials used for foundation construction were procured locally. Cement and reinforcement bars were procured from Indian manufacturers with rigid quality checks in place. Cement storage was made in watertight sheds from where cement was taken to mixing place by closed elevators. Sand used was from the bed of Damodar River (about 160 km away), and it could be used directly because it was clean and matched the specified grain size distribution. The stone aggregates came from natural gravel quarries at Dhalbhumgarh, about 200 km away. The gravel as quarried came with lot of dirt and special arrangements had to be made for washing, and then storing in elevated hoppers.

Concrete was mixed in central batching plants, located on both banks, and then delivered through two cubic yard (1.5 cum) capacity flat-bottomed rectangular skips, having chutes at one end for pouring. Even though it was rather advanced technology for those days, weigh batching was adopted for mixing concrete, in preference to volume batching; the latter is simpler in operation, but the output quality is less dependable as the components are not of uniform consistency (Howarth & Shirley-Smith, 1947).

Each caisson block was served by four 10 ton capacity steam-driven Derrick cranes with 120 ft (36.6 m) jibs. For dredging out the soil from inside the caisson chambers, 60 cubic ft (1.7 cum) capacity grabs were employed. The dredged spoils were removed regularly from site by train bogies moving along rail track arranged by the owner client, the port commissioners, to keep the site clean.

All materials were delivered to the site by rail, and new tracks were laid connecting with the existing tracks for this purpose. Areas were earmarked for storage of construction materials and feeding of materials planned to match the consumption pattern. During the sinking of the caissons, the peak turnaround of 12-ton capacity wagons for removal of dredged spoils reached 30 per day. The peak for supply of construction materials reached 25 in a day.

The caissons for tower foundations were large, 181.5 ft (55.3 m) by 81.5 ft (24.8 m), divided into 21 chambers, 7 by 3 in formation, having 5.73 ft (1.78 m) thick external walls and 5.24 ft (1.6 m) thick internal partition walls. For the convenience of pouring concrete, climbing steel shutters were provided with welded brackets for supporting walkways, which allowed movement of skips for pouring of concrete, and also quick movement of personnel. The gangways that served as working platforms were moved from one location of work to another by Derrick cranes stationed outside the caisson. With heavy concentration of reinforcement bars in the walls, the pouring of concrete had to be done carefully to ensure complete filling and good compaction.

On the Calcutta side, two rail tracks were run between the tower foundations and the anchorage. An elevated gantry structure was built for uninterrupted construction work.

The tower foundations on the Howrah side were located over an abandoned graving dock that had been filled in. The scrap materials and steel structures left behind created problems for sinking of the caisson.

The work on caissons started with the assembling of the cutting edge at the bottom of the monolith (curb steel work) (Figure 6.1).

The ground on the Calcutta side was leveled after removal of all surface debris and underground obstructions like foundations of old buildings. A base was made with broken bricks over which timber sleepers were spread for the laying of curb steelwork.

At the Howrah end, the walls and concrete base slab (of the graving dock) had to be dug out to a depth of 36 ft (10.9 m) below ground level to avoid problems during sinking. As deep excavation was involved, a cofferdam had to

Enclosing the area with steel
sheet piling driven from above
the highest high-water level,
down to the clay

FIGURE 6.1 Assembling of curb steel work for main pier monolith

be built to close the original entry path of the graving dock, to prevent collapse of the bank. The excavated pit was filled back with broken bricks and a level ground prepared for laying of the curb steelwork.

For pneumatic sinking (i.e., sinking with the help of compressed air), the chambers had to be made airtight. Provision for use of airlocks had been envisaged at the design and planning stage and therefore included in the contract conditions. The contractors came prepared with standard "Davis" brand airlocks, which was fitted with shafts in the shape of an "8," and had provisions for separate entry and exit for materials and personnel. Airtight diaphragms were planned for in advance to ensure trouble-free operation with compressed air pressures going up to three times the atmospheric pressure (40 lb./in^2 or 2.81 kg/cm^2) (Ward & Bateson, 1947).

Machined steel castings were riveted on the steel walls of the curb, for seating of the diaphragms on the chambers, and rubber rings fitted on them. To meet exigencies likely to arise during sinking operations, all the chambers were fitted with arrangements for seating of the airtight diaphragms, each of which weighed 23 tons (Howarth & Shirley-Smith, 1947).

CONSTRUCTION WORK: EXECUTION AT SITE

The laying of the well curbs, or cutting edges, started in May 1937, about six months after the start of operations at the site. The extensive and well planned preparatory work ensured the smooth progress of the sinking of the 50 shafts on both banks. While the towers had 21 shafts each, the 4 anchorages accounted for 8 shafts.

The first task was to assemble and fabricate the elaborate curbs that were the same size as the caissons. For the main towers, the curb was a rectangle measuring 181.5 ft (55.3 m) by 81.5 ft (24.8 m) with intermediate partitions to create the 21 chambers of 20.5 ft (6.24 m) square. As pneumatic sinking was being envisaged and the airtight diaphragms were to be fitted during sinking, it was imperative that the chambers were made in the correct size and shape, to be able to accommodate the diaphragms, which, once assembled, were too rigid to be manipulated.

The curb steelwork was assembled on a bed of timber sleepers (with steel packs where necessary) laid on a brick base course, and leveled, before stitching of the separate pieces. The setting out was done using precision theodolites. Longitudinal and transverse alignments of the curb assembly were checked. Rigorous checking was done to verify whether the whole of the curb and the chambers were correctly laid out and correct in size, with all the right angles being checked with steel gauges (Ward & Bateson, 1947).

Bracings provided in the steelwork ensured stability and rigidity to the curbs. Riveting at splices was done after checking the geometry. Wherever corrections were needed, the steelwork was tack bolted, and holes made initially undersized, to be reamed to the proper size prior to riveting. Each square chamber was checked by inserting a precisely made template, so that the airtight diaphragms could properly bear on the machined steel casting prefixed on the walls. All these precautions during fabrication paid rich dividends as no problem was faced in fitting the diagrams.

During actual sinking work, compressed air had to be used in 25 out of the total of 50 chambers and satisfactory airtightness was achieved in the chambers due to the accuracy in the survey and the fabrication

Once the curbs were correctly installed, the sinking operation started by dredging earth from inside the chambers. The Howrah monoliths were sunk through impervious clay strata for almost the entire stretch, down to 87 ft (26.5 m) below ground level, where stiff yellow clay was encountered. This layer had been earlier identified as the founding stratum, having adequate bearing capacity. Sinking was continued for another seven ft (2.13 m) inside the stiff clay layer and bottom plugging of the chambers was taken up progressively after dewatering each chamber. While the work of plugging was being done in one chamber with dredged level at the bottom of the curb, adequate backfilling was maintained in the adjacent shafts, to prevent any trans-chamber movement under the partition walls, with consequent upheaval and disturbance of the founding strata (Figure 6.2).

On the Calcutta side, in the course of dredging, many sunken country boats with cargoes of steel flat bars and reinforcement bars were encountered.

FIGURE 6.2   Progress of monolith wall

They were removed and some were broken up by the heavy grabs and by the weight of the sinking monolith (Ward & Bateson, 1947).

The sinking progressed uniformly and at a depth of 50 ft (15.24 m) below ground level a compact fine sand layer was encountered. Sinking through this 30 ft (9.14 m) sand layer was very even. However, with the dredging progressing below the level of the bottom of cutting edge, there was a sudden drop of the entire monolith at night, when workers were not there. The vibration was felt for quite a distance and some nearby structures, including a temple, suffered damages. The sinking could be restarted after inspection of the monolith and bottom strata. The monolith was finally founded 6 ft (1.82 m) inside a stiff clay layer with the bottom of the curb at 103 ft (31.4 m) below ground level.

The sinking of the monoliths was steady, and the rate of sinking was uniform for all the monoliths. The concreting of the monolith walls had started on both sides in September 1937 and the entire sinking operation, including plugging, was finished by May 1938 on the Howrah side and by November 1938 on the Calcutta side.

The accuracy of sinking and placement of the caissons were phenomenal, with the shift of the foundations from the theoretical location being limited to 3 inches (7.62 cm) – a great achievement for the engineering team.

While the sinking effort generated by the weight of the large monoliths was a contributory factor, the care taken by the construction team achieved such an accuracy by maintaining the level and position of the caissons, all through the sinking duration.

Application of compressed air helped in speeding up the sinking process. Compressed air was used for concreting of the bottom plugs in a dry condition. Quite a few innovative ideas were used in this project by the construction team.

INNOVATIONS

While a seating arrangement for fixing airtight diaphragms were provided inside each chamber, after careful study it was decided that a maximum of four diaphragm sets would be adequate to plug the bottom of the shafts of the large monolith. Out of the four, only two were to be in operation at a given time, as the installation and removal of the equipment took considerable time. Work inside a shaft was continued without interruption until the plugging was over (Figure 6.3).

FIGURE 6.3 Fixing cap for pneumatic sinking

Air pressure was maintained above 40 psi, about three times the atmospheric pressure, and that permitted workers to observe a four-hour work shift. Adequate time was needed for decompression after exit. The capacity of locks for movement of men was small and that slowed down the work. This problem was overcome by rapid decompression in that lock after exit, followed by a second recompression and decompression in another larger chamber fitted with locks for worker movement.

The small size of locks used for removal of excavated spoils and for bringing in concrete for plugging was the other bottleneck. The concrete quantity was much larger than the muck being taken out. This problem was solved by extending these locks vertically such that two cubic yards of concrete could be fed at a time for the plugging operation (Howarth & Shirley-Smith, 1947) (Figure 6.4).

Special care was taken to balance the effect of compressed air in adjacent chambers, while plugging was being done and the cutting edge in the partition wall was exposed fully. Apart from having back filling in the adjacent chambers, the chambers were filled with an adequate height of water to balance the uplift pressure.

On completion of plugging of the entire monolith, the foundations were backfilled with water, such that the pressure underneath was close to the ultimate pressure during service stage. As the erection progressed and dead load increased, water from inside the chambers was pumped out. Thus, the foundations were exposed to full load condition immediately on completion of

FIGURE 6.4 Extended muck lock and bucket for pouring concrete during plugging

plugging operations itself. Adoption of this practice ensured accelerated pre-settlement of the foundations prior to the bridge reaching service condition.

With the completion of the caisson sinking work, the concreting of the reinforced grillage beams, that formed the cap under the tower posts, was taken up. The cap was covered with sloping thin reinforced concrete slabs, for giving the look of a truncated pyramid, matching with the base of the posts. Construction of these slabs needed special collapsible shuttering and careful handling but added to the aesthetics of the giant structure.

The smooth and successful construction of the foundation structures within a comparatively short time span of 18 months set the stage for completion of this massive bridge in less than six years.

FIGURE 6.5   Tower Foundation shaped like truncated pyramid

REFERENCES

Howarth, George Eric & Shirley-Smith, Hubert, *Paper No. 5612, Institution of Civil Engineers – The new Howrah Bridge, Calcutta: Construction*, 1947.

Ward, Arthur M. & Bateson, Ernest, *Paper No. 5601, Institution of Civil Engineers – The New Howrah Bridge, Calcutta: Design of the Structure, Foundations and Approaches*, 1947.

Steel Work: Supply and Fabrication

7

With the completion of the foundation work at the site, the construction of the superstructure commenced immediately. To ensure the continuity of work at the site, planning for the procurement of steel, fabrication of steelwork and manufacturing of special components had to be, and was, done well ahead.

SELECTION OF STEEL

The steelwork in the superstructure was about 26,500 tons and most of it was a high tensile special grade of steel that had a yield stress 35% higher than mild steel to BS 153, which was mostly used for steel bridges then.

The high tensile steel was just coming into use and was not being manufactured by steel plants, and there was no code or specification as yet covering such material. The only bridge project where this comparatively new high strength steel was used was the reconstruction of the Chelsea Bridge in London, across the Thames, between October 1934 and 1937. The success of this innovation at Chelsea encouraged steel designers to use the material, as a significant reduction in quantity of steel requirement could be achieved and thereby savings in cost. For the designers of Howrah Bridge this was a timely godsend. The bridge type selected could now be built within the allocated budget due to savings in steelwork weight by use of high tensile steel and in the cost of the foundations due to the greatly reduced superstructure weight.

The steelmakers (already identified for experimental production) had to be given the full specification of the designed grade of high tensile steel, and this had to be drawn up afresh, as none existed. Plain manganese steel and manganese-chromium steel were in general use, and the specifications of the high tensile steel were drawn by adjusting the percentages of carbon and manganese in those steels to obtain all the required physical properties, especially the tensile strength and the elongation. After a lot of deliberation involving steelmakers, and carrying out a series of tests, the full specification with chemical composition was agreed upon. The final specification was deliberated and agreed with various makers of manganese steel. A series of tests were carried out with identified steelmakers, two from Great Britain and one from India – the finally approved chemical composition and the specified physical properties are indicated in Tables 3.1 and 3.2 (Chapter 3) (Ward & Bateson, 1947). The specification required the steel in main gussets and pin plates to be "normalized" to improve their physical properties.

Extensive testing was done for ensuring the quality of the rivets that were adopted as the main fastener for the bridge structure. While it was very important that the rivets, after driving, filled up the holes completely after cooling down, the rivets also had to have the required strength axially and in shear in the driven condition. The chemical and physical properties of the rivet bars had to be very carefully finalized to ensure their quality, post manufacture (Ward & Bateson, 1947).

The driving of rivets at the site (often in difficult positions with riveters balancing themselves on scaffoldings), needed special skills and excellent coordination among the members of the team, very often passed down from generation to generation of craftsmen. Rivets were heated in a coal-fired flat burner to a desired temperature, assessed only by the exact shade of red color of the hot rivet by the practiced eye of the riveter. If the rivet were hotter or colder it would not serve the purpose. Thus, not only did the rivet have to be heated to the right temperature, but it also had to be transferred to the final location very quickly so that the rivet did not lose its malleability and it could be driven properly. If the rivet was not at the right temperature while driving, the rivet hole would not be properly filled, and a defective rivet would result. Thus, the skill and coordination of the riveting team becomes very important. The required quality could only be maintained by regularly carrying out tests on rivets during construction. A defective rivet would be chipped off carefully, without disturbing the adjacent rivets, and a new rivet driven in place.

As both high tensile and mild steel rivets were used in the bridge, they were made with heads shaped differently – a simple but much-needed precaution to prevent a mix-up and one that ensured use of the right type in the right location.

The three steel plants that had earlier carried out testing of steel (for establishing the final chemical and physical specifications of high tensile steel for fabrication and rivet bars), were ready to supply the steel for the bridge.

Strong representations to the government were made from the Indian industry, and the local manufacturer, that they should supply most of the materials for the construction of the biggest bridge in the country. Tata Steel had already established their credentials by satisfying the quality tests of their steel. To their great relief, the winning bidders, Cleveland Bridge and Engineering Company of Darlington, appreciated the logic of the representation, and agreed to use the steel manufactured in India, as much as possible, provided they could agree on the terms. The matter was resolved satisfactorily with advantages to both sides. Tata Steel was made the agency for supply of about 23,500 tons of steel for the fabrication of the bridge structure from their Jamshedpur plant. The remaining 3,000 tons of steel, which included very wide plates (needed for the lower part of the tower), and the special items like the large diameter pins, bushes and joints, were ordered from steel plants in England. Additionally, the temporary steelwork required for the erection process, the creeper cranes and the various plants deployed for the construction, came from outside India (Howarth & Shirley-Smith, 1947).

Procuring a major part of steel from local sources was a far-reaching decision, which helped in successful and timely execution of the bridge, and commercially proved to be a win-win case for the stakeholders.

The contractors were spared the difficulties of transporting steelwork by sea, infested with U-boats as war had been declared in September 1939. The delivery of steel materials from Tata to the fabricators in good time was ensured, and this was a major factor in the completion of the bridge in a short time. There were considerable savings in cost for the contractor as well.

For the Indian suppliers and the fabricators, it was a great breakthrough and was a valuable experience for future bridge projects.

For the supervision team monitoring the project, the decision reduced risk and provided better control on the activities at the fabrication shops and therefore of erection work.

FABRICATION OF STEELWORK

The fabrication of steel components of this major structure presented a challenge on multiple fronts:

- The locations of the fabrication shops dispersed on both the banks of the river.
- Fabricating very large pieces to very precise dimensions was necessary due to the pre-stressing adopted to reduce/eliminate the secondary stresses during service.

- The need for trial assembly of the components at the shops to ensure accuracy of fit.
- Large compression members had to be end milled so that 50% of the stress could be transferred by direct contract of the butt (the other 50% being transferred by joint rivets). This was as per the designers' decision to avoid huge unwieldy compression joints.

It had been agreed in the contract that Cleveland Bridge & Engineering would entrust the entire steelwork fabrication to the Braithwaite, Burn and Jessop Construction Company Limited (BBJ) a Calcutta-based company. BBJ was a company formed by three large firms, all steel fabricators – the Braithwaite and Company Ltd., located on the eastern bank about 5 km from the site, Burn & Company on the Howrah bank, about 2 km from the site, and Jessop & Company Ltd., 12 km away on the eastern bank. These locations are indicated in the map in Figure 7.1. Fabricated pieces from the three shops had to be taken for the trial assemblies to a fourth shop (Victoria Works) owned by BBJ. Material movement required good planning, as the shops were in busy and crowded areas of the twin city.

These three organizations were registered in India but belonged to British shareholders. They were actually competitors otherwise, but in their keen desire to participate in this rare mega project right in their backyard, had come

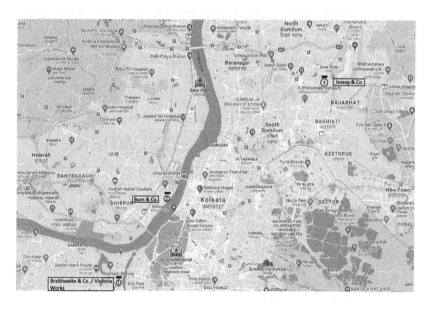

FIGURE 7.1   Locations of fabrication shops

together to form a new company. They had submitted tenders for the full project, appointing capable consultants for alternate designs. To gather the requisite experience, they secured and completed some steel bridges, although very much smaller in size, King George Bridge across River Meghna in East Bengal (now Bangladesh) being one of them.

BBJ's bid to secure the entire work did not succeed but that was not the end of the matter. The idea of a losing such a large project to the UK when materials and expertise were available right at the site of work, did not go down well, not only with BBJ, but also with captains of local industry and chambers of commerce. Their continuous and concerted efforts and lobbying finally produced the desired result. The fabrication work finally came back to Calcutta to BBJ and the steel supply to Tata.

From the beginning of the project, the three members of the BBJ consortium decided to go all out to maintain the very high quality that was required in the bridge steelwork and to deliver as per the schedule laid down by the project management team. To maintain a uniform procedure in the three firms and to prevent their internal differences from affecting the project work, the BBJ board of directors was constituted with the chief executives of the three companies as members.

A unique management system was introduced. Chairmanship of the board was rotated among the board members, to ensure equal participation by the members. A board committee of management formed with the most senior engineer officer of each company as members and chaired by the chief executive of BBJ, monitored the day-to-day activities.

Pre-stressing of the steel girders required the members to be fabricated to precise lengths and laid and assembled for checking that the geometry was correct for the desired joint. For this purpose, an additional shop, Victoria Works, was set up next to Braithwaite's shop at Kidderpore, 5 km from the site. This shop had a suitable shed having adequate length for the laying out of the chords of the truss of the anchor arm or the cantilever arm. An EOT crane with the capacity of handling 60 tons was installed for pre-assembly work. Members fabricated at the other three shops were brought over to Victoria Works for assembly and checked for accuracy of fabrication. The assembled joints were checked with a theodolite for compliance before the final drilling of the rivet holes was done.

All gussets for joints were batch drilled using jigs with hard steel bushes. The same jigs were taken and used in the three shops for corresponding joints so that there was no chance of any misfit in the joints. Chord members were assembled on rectangular cast steel diaphragms, machined all over, fixed on to the members through turned bolts (Howarth & Shirley-Smith, 1947).

All precautions were taken to ensure smooth assembly of joints, as workers had to work at heights exceeding 330 ft (100.6 m) above ground – such that erection work could proceed uninterrupted all through the construction period.

The quality of fabrication was of very high standard. To quote Paper No. 5612, presented at the Institution of Civil Engineers, London, by Messrs. G. E. Howorth and H. S. Smith, senior engineers of the contractor posted at the site:

> The fabrication was very good and suffered in no way from being carried out in four different shops; for example, a careful check on the full length of the lower chord of the anchor arm, comprising eight members, showed that it was correct to within 5/64 inch on an overall length of 325 ft.

It is worth recalling that initially the main contractor was not willing to sublet the fabrication to BBJ, as it was their apprehension that shops in India would not be able to produce the quality of fabrication demanded by a truss super-structure of this size and magnitude.

The following items were manufactured in the UK and shipped to site for use. These items generally required extensive machining work (Howarth & Shirley-Smith, 1947):

- Tension links at the anchors and the anchorage girders, with provision for jacking to compensate for settlements of foundations,
- Pins that were used in various joints in the superstructure, as shown in Figure 4.6 (Chapter 4),
- Hydraulic jacks, of special quality steel, used for closure of the cantilevered arms at the end (Figures 7.2 and 7.3).

FIGURE 7.2   Sliding friction expansion joint

FIGURE 7.3   Jacking at suspended span end

The manufacture and delivery of these items to the site were planned to reach the site well in advance of their installation so as to avoid any hold up in the erection for want of these items.

The bridge work proceeded uninterrupted, despite the destructive World War II raging all around, and this multinational project took shape according to the punishing schedule laid by the project management team.

REFERENCES

Howarth, George Eric & Shirley-Smith, Hubert, *Paper No. 5612, Institution of Civil Engineers – The New Howrah Bridge, Calcutta: Construction*, 1947.

Ward, Arthur M. & Bateson, Ernest, *Paper No. 5601, Institution of Civil Engineers – The New Howrah Bridge, Calcutta: Design of the Structure, Foundations and Approaches*, 1947.

Erection of Steelwork

8

The construction of the superstructure of this 1500 ft (457.2 m) span bridge was the most difficult challenge for the engineers and the planning for the same started from the design stage itself. For a long span bridge, the designer has to formulate the design scheme for the construction stage and has to check for the same along with the serviceability stage. Howrah Bridge was no different.

The conditions of the tender allowed the bidders to formulate their own erection scheme, though the bid document included the scheme of erection conceptualized by the designers while working on the design of the superstructure (Ward & Bateson, 1947).

Since the main structure on either bank comprised an anchor arm and a cantilever arm, cantilever erection with a crane moving on the top chord was the automatic choice for that part. While there was no deck structure for support at the erection stage and the live load was absent, the cantilever structure had to bear the load imposed by the creeper crane that was to do the erection by moving along the top chord, the slope of which kept on changing, complying with the final profile of the bridge.

The suspended span, which was supported on the tip of the cantilever arms from both sides, could be erected by hoisting from the river with the help of floating cranes, or alternatively cantilevered out from the cantilever arms at both ends. After thoroughly examining the implications of both the alternatives, the designers chose the scheme involving continuation of the cantilevering process until the closure at midspan. The chosen scheme required strengthening of the end chords of the suspended span, as during the cantilever process, the top chord of the suspended span was subjected to tension and the bottom chord to compression, unlike the service condition when it is the otherway around. The end chords of the suspended span needed to redesigned for this erection condition.

Further, the deflection at the tip of the extended cantilever and the axial deformation of the chords were enhanced in this scheme of work. As a result, provisions had to be made to reverse the effects prior to the closure stage, such that the bridge when completed did not retain any locked-in stress arising from the erection process.

The erection by lifting of the assembled suspended span, with the help of floating cranes, was avoided as arranging such heavy-duty cranes would have added to the procurement time and the cost, whereas the creeper crane moving on the top chord was already available for use (Howarth & Shirley-Smith, 1947).

Even though offered the freedom to change the erection scheme, none of the bidders came forward with a better and more construction savvy scheme.

ERECTION EQUIPMENT

The main equipment deployed for the erection of the bridge at either bank were:

- 60-ton capacity derrick crane, mounted on the ground, stationery at the approaches, used mainly for unloading of fabricated steelwork.
- 60-ton capacity portable three-legged stand with a 60-ton tackle fitted at the top, used for handling and unloading heavy pieces at locations away from yard and for transferring to different tracks (Howarth & Shirley-Smith, 1947).
- 60-ton capacity creeper cranes with twin derricking booms, capable of unlimited slewing, and moving along the top chord of the truss. The main hoist could lift 60 tons at the radius of 40 ft (12.2 m) and an additional auxiliary hoist was provided for lifting 20 tons at a radius of 90 ft (27.4 m). The crane undercarriage was wheeled and designed to move on pair of rails riveted on top of removable tracks, formed with heavy box girders fixed on top chords with the help of temporary stools. The track could be lifted from the rear after the crane moved forward and re-laid in front, to enable uninterrupted movement.
- The undercarriage was carried on a triangular cradle while erecting the anchor span because of steep slope of the top chord.

This equipment was specially designed to be capable of hauling itself up the anchor arm, having slopes of around 30 degrees, and then cross over on the other side of the tower, for carrying out the cantilever erection up to the center line of the bridge, prior to closure.

The crane had to lower itself, with the changing slope of the cantilever arm from 22 degree to 9 degrees, and then move along the horizontal top chord of the suspended span. For moving along changing slopes, the upper part of the crane was pivoted on the undercarriage and the axis was maintained at vertical at all times by the use of screw levelling gears provided between the back of the crane and the carriage. A levelling mechanism was provided to keep the mast of each crane vertical, and a pendulum was used to monitor the verticality (Howarth & Shirley-Smith, 1947; Ward & Bateson, 1947).

The crane was hauled with the help of two winches mounted on the under-carriage and operated by pulling on heavy wire rope tackles that led to a cross head bolted on the end of the track at rear, through groups of sheaves on the undercarriage. While hauling up on the anchor span a force of 450 tons was necessary for starting and 419 tons for moving. The weight of the crane was 610 tons and the undercarriage weighed 95 tons.

These unique pieces of custom-built crane units were designed by the contractors, Cleveland Bridge in collaboration with Messrs. Wellman, Smith and Owen. The crane was operated with an electric supply at 400/440 volts, 50 cycles.

Safety precautions were given the highest priority. Before lifting loads, the cranes were locked to the track, and a protective electrical system ensured that the hoisting motors could not work unless the locking was done. Brakes were provided so that the crane would not slip down even when on the steep-est slope. Cut-out safety switches were provided for controlling all operations and signals by means of warning lights and bells, and became operational whenever the loads on cranes got close to their capacity. Independent elec-tromagnetic foot brakes were provided in the hoisting system and telephonic communication maintained with the control room at the ground.

All these precautionary steps paid rich dividends, and not a single mal-function affected the crane operations.

The senior engineer in charge of the creeper crane design, erection and operation, Dr. H. P. Budgen ascribes the operation skill of the workmen in his recorded comments:

"Indian drivers operated the crane movements, and tribute is paid to their care, enthusiasm, and ability"

(Discussion on the new Howrah Bridge, 1947)

The site riveting work was done by using pneumatic tools. The air compressors were also used to blow air inside the boxed towers, for workers to work without being affected by the fumes generated by the paints burned during riveting of joints.

ERECTION OF STEELWORK
IN ANCHORAGES

The anchorages at the four corners housed the mechanism for controlling the level of the completed truss, which could become necessary to obviate the effects of unplanned settlement of the tower foundations, which were ultimately resting on clay layer. This mechanism could also be used for the final closure of the trusses at the center joint at the end of cantilevering operation.

The entire uplift force at each anchorage was in the order of 6,500 tons and the same had to be transferred to the foundation block through the link mechanism. The transfer of the load was affected through the three main steel girders located at the bottom of the anchor shaft, connected to the assembly of three pin-connected vertical links with double web plates, hung from the bottom corner of the anchor truss, about 90 ft (27.4 m) above the steel girders.

The steel girders transferred the entire uplift force to a set of eight grillage beams on top, their ends embedded into the concrete walls of the anchorage block, thereby harnessing the dead weight of the block and the frictional force from the surrounding soil mass against uplift.

To allow adjustment in the levels of the anchor truss ends, an arrangement was made for including 16 hydraulic jacks between the steel girders and the grillage beams and inserting adjustable packs at 24 contact points between them (Ward & Bateson, 1947) (Figure 8.1).

With this mechanism in position, it was possible to vary the levels at the ends of the anchor trusses and compensate for any effect of settlement of tower foundations and also have control on the level at the tip of the cantilever arms. This mechanism is still in operational condition.

The bottom links were 43 ft (13.1 m) long and each weighed 45 tons. Lifting them out of the wagons and placing them inside was a challenging task and could be done by using the portable derrick cranes. The anchor girders and the links had to be adjusted properly to bring them to design level before the connecting pins were introduced. Jacks were put in position on diaphragms between grillage girders. The packings and the grillage beams were erected on the anchor girders and then the concrete of the wall above cast, to ensure that the uplift forces were directly transferred without any slack. While sinking the anchorage foundation, this part of the wall was blocked with brickwork, which was removed prior to concreting the wall above the grillage beams.

This arrangement of jacking the ends provided the flexibility required when working with huge loads on clay as the foundation material. It came in handy when the anchorage blocks at the Howrah end settled by 11/16th of an inch (18 mm) due to the heavy load of the creeper crane put on the anchor truss,

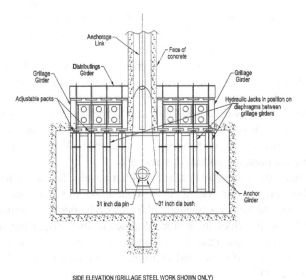

SIDE ELEVATION (GRILLAGE STEEL WORK SHOWN ONLY)

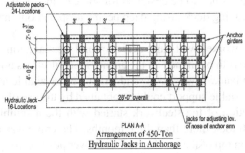

PLAN A-A
Arrangement of 450-Ton
Hydraulic Jacks in Anchorage

FIGURE 8.1 Arrangement for adjustment by hydraulic jacks operating on the grillage beam

foundation block. Problem was solved by inserting packing plates of 3/4th inch (19 mm), after jacking down the anchor girders, against the grillage beams (Howarth & Shirley-Smith, 1947).

ERECTION OF ANCHOR SPAN STEELWORK

The erection was started from the anchor spans at either bank, and work at the Howrah bank could start about three months earlier than planned as the foundation work was completed earlier with fewer problems, than with the sinking of caissons at the Calcutta end.

For erection of the anchor span steelwork, temporary trestle supports were constructed. As the area under the anchor spans was close to the river bank, special provisions were made to avoid settlement of the trestle supports. The trestles, which were located at each main panel joint and subpanel, had to carry weights ranging from 200 tons to 900 tons per truss, from the self-weight of steelwork and the shared load of the creeper crane.

On the Howrah side the soil strata did not permit supporting the loads on piles. Further, the old graving dock that occupied the area earlier had been excavated and backfilled with heterogeneous materials. These posed problems in providing temporary supports for steelwork.

Ultimately, the contractors came up with a novel solution – a foundation based on buoyancy. A thin walled box without a bottom was constructed with reinforced concrete and sunk to about 20 ft (6.1 m) by open excavation. It was then provided with a floor of reinforced concrete slab, displacing soil mass having 50% more weight than the weight to be imposed. The base slab had to be designed to transfer the load on to the ground underneath (Howarth & Shirley-Smith, 1947).

Special foundations had to be designed for some of the trestles that came on the edge of the graving dock, using deep beams to spread out the load over larger areas.

The foundations worked well during erection and the maximum settlement that occurred was 11/16th of an inch, in the anchor block itself, as noted earlier.

On the Calcutta bank, where good sand strata were available at depths of about 50 ft (15.2 m), stable support could be obtained by the use of vibro piles.

The actual erection of the steelwork started with the assembling of the creeper crane, which took more than three months. The crane was erected on an assembled track on top of the trestles near the tower position for placing the main bearings at the tower location. The lower chord of the anchor span next to the tower was then erected. The crane then started moving back towards the anchorage, on its way erecting the bottom chords one by one.

The chords were erected on steel packing supported on the trestles so that on completion of the truss it could be put in its cambered profile by jacking the chords at the trestle points as necessary by use of hydraulic jacks.

Once the crane reached the anchorage, it was put on the triangular-profiled cradle, on a temporary steel ramp behind the anchorage, for its eventual climb on the top chord of the anchor truss. From there it could carry on erecting the rest of the members as it proceeded towards the tower (Figures 8.2 and 8.3).

The towers were also erected by the creeper crane from its position on the top chord of the anchor span. The saddles on top of the tower were 24 ft (7.3 m) long and 21.5 ft (6.5 m) high and were made up of three built-up webs interconnected by diaphragms. The saddles had four machined holes (with diameters of 30 in and 18 in) for the pins connecting the chords and the diagonals.

FIGURE 8.2   Creeper crane on anchor truss

FIGURE 8.3   Creeper crane crossing over tower top

The pins could be erected without much difficulty, despite their weight of up to 5 tons per set, and the need to pass them through five plies of steelwork in the saddle and member webs. After completion, the tower on the Howrah side showed its legs to be free from twist and the top of the tower was within 3/16 in (4.5 mm) of its theoretical position. This was yet another testimony to the quality of fabrication of the steelwork (Howarth & Shirley-Smith, 1947).

Because of the heavy weight of the crane, it was necessary to introduce temporary steel members for strengthening the chord members. A temporary top lateral system and sway bracings also had to be introduced, as the design did not provide permanent lateral bracings to the top chords, which were only tension members in their service condition.

The time lag between the erection work at the Howrah and Calcutta bank allowed the temporary steelworks used in the trestles, ramp and strengthening members to be transferred to the Calcutta end after their use in the Howrah end. The total weight of such temporary steelwork was 2,500 tons, and their reuse offered significant economy to the contractor.

Because of the prestressing of the truss, the upper chord of the anchor arm was fabricated about 3.75 in (9.5 cm) shorter and the tower fabricated 1.75 in (4.5 cm) longer. To make the closing connection, it became necessary to pull back the top of the tower by 13 in (32 cm) and the splices in the chords and diagonals were completed after the tower was jacked at the bottom (Figures 8.4 and 8.5).

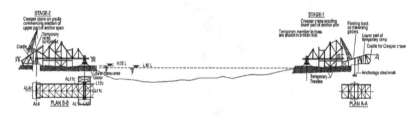

FIGURE 8.4 Erection stages 1 and 2

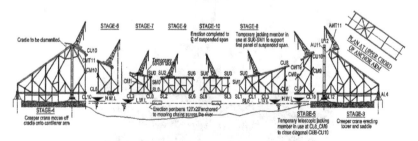

FIGURE 8.5 Erection stages 3 to 10

ERECTION OF THE CANTILEVER ARM

The crane had to cross over the tower to commence erection of the cantilever arm. The first panel of the cantilever arm was erected with the help of the creeper crane and then the tracks were assembled on the cantilever arm. A specially made bridge piece was fitted between the cradle on the anchor span and the track made on the cantilever span. The cross heads were dismantled and re-erected in their new position. The hauling ropes were moved onto the new position and the crane lowered off the cradle into the new position on the cantilever arm for commencing erection work. The cradle used for erection of the anchor arm was removed and dismantled (Figure 8.3).

The first two panels of the cantilever arm were supported on trestles and pre-stressing done as needed to complete the connection, by bending and displacement of the members. Jacking was also needed from the top of the trestles to help in closure of the joints (Howarth & Shirley-Smith, 1947) (Figure 8.6).

For subsequent panels, trestles were not available and the fitting in of the pre-stressing members had to be done by other methods. The tower top was inclined backwards at the initial stage and the forces required to make the members match with the holes in the gussets, were larger. Temporary members were introduced to move the joints and the members. The creeper crane helped in introducing the required pull with the help of the wire rope tackles. Some of the permanent members were provided with an in-built jacking facility (planned earlier) and some members were manufactured with telescopic movement and with provision for fixing hydraulic jacks. Bottom chords could be lifted by calculated amount to make the joints.

With the progress of the cantilever erection, the stresses in the anchor truss members approached normal values and the tower moved forward towards its normal position. The force required for prestressing the members reduced progressively in the cantilever arm.

Once the cantilevering was completed, the center line of the truss was checked and the lateral deviations from theoretical position were found to be of the order of 1/8th in to 5/16th in (3 mm to 8 mm).

ERECTION OF SUSPENDED SPAN

The suspended span was designed to be simply supported between the tips of the cantilever arms, hung on four vertical links at the end of the cantilever arm.

FIGURE 8.6   Commencement of erection of suspended span

The span was restrained longitudinally by locating gears on either side of the central universal joint, and laterally by the universal joints at the center and the sliding bearings at the ends of the cantilever arm.

Both at the lower chord and upper chord joints, at the meeting point between the cantilever arm and suspended span, 800-ton hydraulic and 450-ton screw jacks were provided that could move the joints forward or backward at all the four corners at each end of the suspended span. With the help of these jacks the chords could be advanced towards each other. The cross struts at the junction point were assembled on pontoons along with center gussets, universal joints and jacking steelwork, weighing 90 tons all together, were lifted by two creeper cranes working in tandem (Figure 8.7).

The erection of the suspended span could be started with a temporary adjustable member used between the top corner of the cantilever arm and the

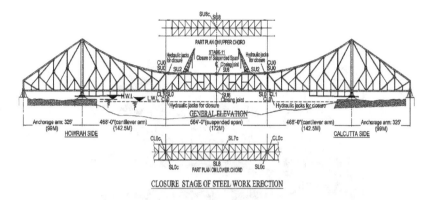

FIGURE 8.7   Closure stage of truss erection

first panel of the bottom chord of the suspended span, and the same support had to continue until the span erection continued up to the mid span. At this stage, the lifts became lighter, and the creeper cranes were lightened by reducing the kentledges on the crane. This helped reduce the forces on the extended cantilever arm.

Erection of the suspended span was considered complete after the upper and lower chord forces were transferred to the screw jacks.

CLOSURE OF THE SUSPENDED SPAN

Hydraulic piping was laid to connect the jacks, pumps and valves required for operating the jacks and the corners of the chords could be moved out or inwards, either independently or in a synchronized manner. Trial were conducted in advance on the hydraulic system such that no failure happened during the closure operation. Squads of people for working on the eight corner joints were trained and rehearsed on the operation such that no mishandling took place during the time-bound closure operation (Figure 8.8).

The final operation was to be in three parts – closure of the lower chord, closure of the top chord and swinging the suspended span in final position. All three operations were carefully rehearsed, so that all the deputed staff knew exactly their duties at each stage. Completion of the operation in the shortest possible time was given the utmost importance as the chords would get distorted due to the heat from the sun with the progress of the day. It was recorded

FIGURE 8.8   Creeper cranes approaching for closure

that after the chords of the half spans had been advanced towards each other, the gap at the lower chord at the minimum prevailing temperature was 4.5 in (114 mm) in the lower chord and 5 in (127 mm) in the top chord (Howarth & Shirley-Smith, 1947).

The faces of the top chord joint were parallel, but the bottom chord joint had the gap at the upper surface 5/64 in (2 mm) more than at the bottom.

Ultimately, the closure operation began at the break of daylight on December 30, 1941, after checking with the weather department that the weather would remain stable.

The lower chord gap was closed by jacking and the joint was bolted with increased jacking force of 200 tons. Clouds were there until 9 a.m. and the additional gap in the top chord could be closed with the help of the jacks, working from both ends. The sun got stronger and distortion of the chords started.

The swinging operation started at about midday and with controlled operation of the jacks, by reducing the pressures slowly, the operation could be completed by 1 o'clock.

It was found that the lower chord of the suspended span had increased by 4 in (102 mm) and the top chord shortened by the same amount. The level of the suspended span was still 28.75 in (730 mm) above the calculated geometric level (Figure 8.9).

The jacks and the temporary jacking steelwork were dismantled and removed. The creeper cranes were moved to the center and dismantled. The bridge structure was completed from the engineers' point of view, as the deck structure, over which vehicles and men would cross, was merely an additional load for the bridge truss to carry.

FIGURE 8.9   Closure of truss

The tension-filled closure operation was complete without any hitch – a tribute to the meticulous planning and rehearsals carried out in advance.

ERECTION OF DECK STRUCTURE

The deck steelwork was mostly erected with the help of hand winches and feeding of steelwork was partly by pontoons moved by tugs and placed directly below the truss.

The hangers, cross girders and the stringers were erected first and then the floor beams, troughs and curbs were fed along the track laid on the deck and skidded into position. The entire deck steelwork was put in position less than three months after the closure date (Figure 8.10).

FIGURE 8.10   Erection of deck steelwork

SITE RIVETING

The riveting work was an important site operation for the safety and longevity of the span. The site rivets were 1 in (25 mm) diameter and the grips varied from 2 in (51 mm) to 6 in (152 mm). The heads of the high tensile steel rivets were by choice made conical, to prevent mixing them up with mild steel rivets.

Riveting of the tower joints were done by using steel staging, but for other positions, timber plank staging was used. The steel staging was handled by the creeper crane, and the timber planks by manual effort, using lifting tackles.

For riveting work inside the tower shafts, that were about 3 ft (91 cm) square, and 270 ft (82.3 m) high, ventilation arrangements were done to remove the fumes created by burning of paints. The riveting work progressed without any hitch and was of good quality. This was recorded by the senior managers of the contractor, G. E. Howorth and H. Shirley-Smith, chroniclers of the article presented in Institution of Civil Engineers, London:

> The riveting was carried out by gangs of Punjabis or Pathans under an English foreman; the work was first-class and replacement of rivets seldom amounted to 1 percent. The average output per squad of four men per shift was about 500 rivets.
>
> (Howarth & Shirley-Smith, 1947)

The significance of the above statement can be better understood from the fact that the experience of the present author had earlier convinced him that the maximum production from a rivet gang of five people can at most be 400 rivets in a ten-hour shift – and that too in an easily accessible steel girder bridge!

The erection of the entire steelwork for the bridge was completed on March 1942, 28 months from the start of anchorage steelwork erection in November 1938. This was an achievement, during the war months with Japanese bombers hanging around to bomb the bridge, giving the construction engineers goose-pimples! The bridge, however, could not be opened to traffic even though completed in all respects by August 1942.

Those were the troubled times of history. The Indian political scenario had become restive with the declaration of non-cooperation with the government at all levels, in support of the demand for total freedom from colonial rules. Bengal was having the Great Famine that left many dying of hunger on the streets. The war theater in the south-eastern front was in gloom for the allied forces and curfews were frequently imposed. The completion of such a

FIGURE 8.11   Baptism of Indian workers in dizzy heights

showcase project deserved a ceremonial inauguration, but the environment was not conducive.

Finally, the bridge was declared open after running a single tram car across the bridge at midnight on a date in February 1943, unknown to the city dwellers. The provisions for lighting and traffic control could not be completed. Temporary signs were erected, and, without any traffic control, various streams of traffic started operating on the bridge freely without any control from the traffic police.

The bridge had become an attraction for local people even before it was completed. Huge crowds of bystanders could be seen on the banks. They were struck by the workers moving along dizzy heights nonchalantly – certainly, a greater attraction than the performing circus units that used to visit the city (Figure 8.11).

Construction of the huge structure in front of the curious eyes of the citizens made it an instant hit and the love affair still continues after 78 years – the bond has only increased after heavy traffic started using the cable-stayed bridges completed in 1994 and the pedestrian traffic soared across the bridge.

REFERENCES

Discussion on the new Howrah bridge – *Paper Nos. 5601 & 5612, Institution of Civil Engineers*, 1947.

Howarth, George Eric & Shirley-Smith, Hubert, *Paper No. 5612, Institution of Civil Engineers – The new Howrah Bridge, Calcutta: Construction*, 1947.

Ward, Arthur M. & Bateson, Ernest, *Paper No. 5601, Institution of Civil Engineers – The New Howrah Bridge, Calcutta: Design of the Structure, Foundations and Approaches*, 1947.

An Epilogue

9

There are other aspects of the bridge that merit examination.

THE BRIDGE AS CALCUTTA'S ICON

Now that we know the details of the evolution, design, fabrication and construction of Howrah Bridge, we are faced with a question – why had this bridge become and continues to be the icon of Calcutta? There are other worthy claimants for it – the aesthetically pleasing Victoria Memorial, which will celebrate its centenary in 2021, or perhaps the other younger bridge a kilometer downstream on the Hooghly, the sleek and much lighter Vidyasagar Setu, opened 50 years later than the Howrah bridge? We can only guess the possible reasons.

Was it the massive structure, which was functionally such a boon and suddenly gave freedom of movement for goods and people, removing the constraints caused by the pontoon bridge and the ferries? Or was it the pride of having a great bridge, the like of which there were only two in the world then? Or was it the fact that Indian steel and Indian workmanship had come of age and could stand comparison with the best in the world? Perhaps the construction happening against the socio-political background in the turbulent and dark days before the country got its freedom, was the reason. Whatever the reason, the almost 80-year-old Howrah Bridge and the 350-year-old Calcutta city are inseparably connected.

RELIEF FOR THE HOWRAH-CALCUTTA TRAFFIC

The old pontoon bridge had given excellent service, but it just could not serve the huge growth of goods and pedestrian traffic. Crossing the river took inestimable time as the bridge was always brimming with vehicles and pedestrians.

81

From February 1943 onwards, the new Howrah bridge provided an enormous relief. With many lanes, easy approach slopes and a tramway being available, river crossing was no longer a torment. There were all-round commercial and social benefits. It was the same for pedestrians, though within a span of a decade the footpaths were teeming with people.

It has been estimated that the number of pedestrians using the bridge now is anywhere from 200,000 to 1 million. The designers had thoughtfully provided extra wide footpaths but that could not match the pedestrian growth. Robust guard rails have ensured that pedestrians have not been pushed into the river or on to the carriageway (Figure 9.1).

DIDACTIC AND ENGINEERING FEATURES

The bridge was a state-of-the-art structure in its time, and even after the passage of all these years, still has features noteworthy for today's professionals.

FIGURE 9.1   Pedestrians jostling to cross the Bridge

It was the third longest span balanced cantilever bridge, after the formidable Firth of Forth bridge near Edinburgh and the Quebec bridge in Canada, but it presented more problems due to its location in a heavily built-up urban environment.

The construction of Howrah Bridge included several firsts and examples of good engineering for future bridge engineers to emulate. This was the first bridge to have multi-chamber rectangular caissons as the foundations, a design that ensured good control of tilt and shift in treacherous heterogeneous alluvial substrata, sunk with the help of controlled dredging and other means. This was also the first bridge to use pneumatic sinking in India, to enable sinking through "boiling" fine sand in a water-bearing stratum, which the bore charts had indicated at the location of the Calcutta caissons.

This was the longest single span in the country, until it was matched by the cable stayed Vidyasagar Setu, a kilometer downstream, which was constructed in the last decades of the twentieth century. Before Howrah Bridge, the longest single span in the country was that of the other bridge across Hooghly about 40 km upstream, i.e., the Jubilee Bridge with span of 165 m.

Two innovations helped the adoption of a cantilever truss bridge of such a long span, and it therefore could be constructed within available budget. One was the of use of high tensile steel for the first time in a major bridge. This reduced the weight of steelwork substantially in the cantilever arm thus effecting large all-round savings in steelwork arm foundations.

The other was pre-stressing of the members to reduce/eliminate secondary stresses, which contributed to savings in steelwork and added to the load-carrying capacity of the bridge. These two innovations were very successful, and these practices have now become the norm in all long-span steel truss bridges across the world.

Nearly 80 years of trouble-free service of the bridge is undoubtedly due to the forethought of the designers in providing for all foreseeable problems and the quality of construction.

In the design stage, erection stage requirements were calculated, and jacking arrangements were provided for adjustment. Corrective action against deflection of the tip of the cantilever and changes required in the length of chords during erection were computed by the designers. Effective arrangement of jacking the span from inside the anchor foundation (part of it to cater against possible settlements) and provision of jacking/pulling arrangements in the chords reflects the minute care the designers exercised and this ensured a perfect closure, at the end of erection, of the bridge possible

Disproving the common notion that engineers are blind to aesthetics, the designers not only took care to modify the initial designs to give a better elevation to Howrah Bridge, as it was a highly visible structure located in the heart of the city, but even the rivet layout in the large joints have been carefully detailed with an eye on symmetry and appearance.

IMPACT ON LOCAL INDUSTRY

A word on the prevailing political situation in the later half of 1930s is relevant. War was imminent in Europe. The British government was becoming aware that it would have to grant India independence in the very near future. The Indian public were equally aware of this. The situation was volatile. When the tender for Howrah Bridge was floated, there was a strong feeling in India that there should be a strong local participation as there was top-quality workmanship available, though no such major work had been executed by the Indian industry. When, after tender evaluation, Cleveland was about to be awarded the work, there was very strong pressure from British owners of Indian fabrication companies as well as the steelmaker Tata that there should be substantial participation of local industry in the construction of the bridge. As has been mentioned earlier, Cleveland was given the main contract and they subcontracted the steelwork to the local combine BBJ, and Tata supplied almost all the steel. The quality of material and workmanship had been tested and found to be equal to the best available in Britain. The cement and other materials used were also Indian.

Though the consultants and the principal contractor, Cleveland, provided the top managerial staff, the entire supervisory staff and workmen were local. The quality of Tata's steel, the skill of the local workmen and the excellent quality of workmanship have been well documented in the published papers on Howrah Bridge by the top design and construction engineers. That the bridge was finished as planned in spite of an ongoing world war is as much a tribute to the management as to the local workers.

LOCAL STEELWORK INDUSTRY COMES OF AGE

Though the local steelworkers (in steel plants and in steel construction) were very skilled, it was entirely a different situation that in the construction of Howrah Bridge, they were up there with the best in the world and the skills were recognized by the top engineers of Britain. This gave the industry absolute confidence in their abilities, which had now been tested and found to be as good as the best.

For the construction of the 1,829 m long road cum rail bridge that came up over River Ganges at Mokameh, Bihar, a decade or so later, BBJ was more or less the automatic choice for the steelwork and Hindustan Construction Company for the foundations, both companies having established excellent credentials in the Howrah Bridge construction.

This was followed by the construction of another road cum rail bridge, this time across the mighty Brahmaputra River near Guwahati in Assam, in North East India, where the same duo did the steelwork and foundation circa 1960.

In fact, most of the major bridges in India subsequent to Howrah Bridge have been designed and constructed with local engineering expertise. The fabrication shops in Calcutta that manufactured steelwork for Howrah Bridge, supplied steelwork for bridges all over the country, one of the important ones being Vidyasagar Setu. The cable-stayed bridge across the river just a kilometer downstream of the Howrah bridge was opened about 50 years after Howrah Bridge. Construction industry experts across the country acknowledge the contribution of the Howrah bridge to the growth of the construction capability in India after independence.

(On a personal note, two of the steel erection workers who had their baptism in the Howrah Bridge construction, Omeruddin Khan and Harbans Singh, served as erection foremen for many major bridges, and the author had the opportunity of seeing them in action in the construction of some important projects, like the restoration of Hardinge Bridge in Bangladesh, the second Hooghly bridge i.e., Vidyasagar Setu. They rendered yeoman service.)

MAINTENANCE AND PRESENT CONDITION OF THE BRIDGE

Another happy feature is that the maintenance of the bridge was assured of funds as the Howrah Bridge (Amendment) Act of 1935 provided for regular earnings for the bridge as a surcharge on the municipal tax paid by citizens of Calcutta and Howrah, and also on all the train tickets purchased by travelers alighting at Howrah.

A recent inspection and testing program carried out on the Howrah Bridge established that the main truss is in perfectly good shape and did not exhibit any signs of distress or damage. The deck structure had suffered some visible

FIGURE 9.2 Night Illumination of Howrah Bridge

damages due to (some misuse by) the huge pedestrian traffic and corrective measures have been introduced.

Analysis of the design of the superstructure and superstructure revealed that despite major changes in code provision for loads to be carried on the bridge, this bridge still has some reserve capacity left both for the permissible stresses in steelwork and the bearing pressure at founding level. This fact, together with the quality of steel having larger copper content than used now, perhaps gives this bridge the good health that it enjoys even after intense use for 78 years.

The bridge looks like it will be able to provide many more years of useful service, thanks to the care bestowed on it by the port and city authorities. Calcuttans are extremely proud of Howrah Bridge and love to display it as their icon and time has only increased that attachment.

The importance of the bridge was highlighted by the prime minister of India, who unveiled the Dynamic Architectural Illumination of Howrah Bridge in January 2020 as a part of 150 years of the Calcutta Port Trust (Figure 9.2).

In conclusion, let us wish good health and long life to this outstanding structure, a monument to the outstanding work of the designers, the construction firms, dedication and unremitting toil of all the workmen.

Index

Page numbers in *italics* refer to figures and those in **bold** refer to tables.